Lecture Notes in Mathematics

A collection of informal reports and seminars
Edited by A. Dold, Heidelberg and B. Eckmann, Zürich

219

Norman L. Alling
The University of Rochester, Rochester, NY/USA

Newcomb Greenleaf
The University of Texas, Austin, TX/USA

Foundations of the Theory of Klein Surfaces

Springer-Verlag
Berlin · Heidelberg · New York 1971

AMS Subject Classifications (1970): 14H05, 14J25, 30A46

ISBN 3-540-05577-0 Springer-Verlag Berlin · Heidelberg · New York
ISBN 0-387-05577-0 Springer-Verlag New York · Heidelberg · Berlin

Offsetdruck: Julius Beltz, Hemsbach

INTRODUCTION

It has long been known that the category Ⓢ of compact Riemann surfaces and non-constant analytic maps, and the category Ⓒ of complex-algebraic function fields and complex isomorphisms are, via two contravariant functions, coequivalent; thus an analytic theory and an algebraic theory are tied together.

While investigating several Banach algebras on compact Riemann surfaces $\mathfrak{X}$ with non-empty boundary ∂X ($[A_2]$, $[A_3]$, $[A_4]$), the first author posed the following question for himself: what is the simplest algebraic object which can be associated with $\mathfrak{X}$, from which $\mathfrak{X}$ can be recovered? The answer seems to be the following: let $E(\mathfrak{X})$ be the field of all functions f meromorphic on $\mathfrak{X}$ such that $f(\partial X) \subset R \cup \{\infty\}$. This field is an algebraic function field in one variable over the reals. It is natural then to ask the converse question: given such a field E, is there a compact Riemann surface $\mathfrak{X}$ (possibly with boundary) such that $E = E(\mathfrak{X})$? The answer to this question, interestingly, is no. The following field, long known to algebraic geometers, supplies a counter-example to such a conjecture: let $E = R(x,y)$, where $x^2 + y^2 = -1$. The present collaboration began at this juncture.

Let Ⓡ be the category of all real-algebraic function fields and all real-linear isomorphisms. Given such a field E, the algebraic geometers have long known how to associate a curve X with E; for example let $X \equiv \{\mathfrak{O} : \mathfrak{O}$ a valuation ring of E over $R\}$. The usual topology put on such a curve is the Zariski topology in which the proper closed sets are the finite sets. Such a topology does not utilize the topology on R and does not render X a manifold. It is relatively easy to define another topology on X, utilizing the topology of R, under which X is a compact surface (possibly with boundary). At this point in our investigations, we hypothesized that algebraic "functions" on such a manifold must be "analytic," in some sense. Our task was then to find the correct "analytic" structure on X.

In the example given above, X is the real projective plane; thus non-orientable X arise. Since conformal maps are orientation-preserving, the definition of "analytic" structure needed to be relaxed so that non-orientability could occur. The way out of this is to allow the transition functions t to be analytic (i.e., $\partial t/\partial \bar{z} = 0$) or anti-analytic (i.e., $\partial t/\partial z = 0$): that is <u>dianalytic</u>. Having roughed out this dianalytic theory, we discovered that Schiffer and Spencer [SS] had, not surprisingly, done it before us; but -- being less interested in the compact case -- had not come to a full realization of the algebraic consequences of this dianalytic theory. Further, they do not define morphisms and functions directly for non-orientable surfaces. Going back to an old tradition [B], we have chosen to call surfaces with dianalytic structure Klein surfaces. Much of our Chapter 1 lays the foundations for the intrinsic study of morphisms, functions, and differentials on Klein surfaces.

A brief summary of the contents is as follows: Sections 1.1 - 1.4 contain the basic definitions of Klein surfaces, meromorphic functions, and morphisms. The main novelty here is that if $f : \mathfrak{X} \longrightarrow \mathfrak{Y}$ is a morphism then X (the underlying space of $\mathfrak{X}$) may be folded over ∂Y. Sections 1.5 - 1.9 deal with methods of putting dianalytic structure on surfaces: particularly lifting and descending dianalytic structure under covering maps. It is shown that every compact surface can carry dianalytic structure. The dianalytic structures on the disc, real projective plane, annulus, Klein bottle, and Möbius strip are classified. Section 1.10 treats the theory of differentials and their integration on Klein surfaces, while 1.11 deals with automorphisms of Klein surfaces. Most of Chapter 1 is devoted to the general theory of Klein surfaces; in Chapter 2 we deal with the compact case. The main theorem is the following: the category of compact Klein surfaces and the category of algebraic function fields in one variable over R are co-equivalent. That is, compact Klein surfaces can be considered as non-singular algebraic curves over the reals. Various applications of this theorem are given. A research announcement summarizing most of the results of this monograph has appeared [AG].

Our aim in writing these lecture notes is to present a fairly complete account of a recently re-vitalized theory, accessible to non-experts. This is not to say that no background is required; it is. On the analytic side some knowledge of the theory of Riemann surfaces will be an enormous help to the reader. On the algebraic side, some knowledge of algebraic curves and their function fields is required to appreciate Chapter 2. Even with these pre-requisites, we have tried to ease the task of the reader not steeped in these classical theories by supplying some review of known facts. For example, the first section of Chapter 2 is entirely review.

The history of this subject is long, going back to Klein's 1882 Monograph on Riemann's theory of Abelian integrals [K], for -- in the closing pages -- he considers the group of conformal maps of the Klein bottle, and other non-orientable surfaces. No definition, acceptable to a modern reader, seems to have been known before the appearance, in 1913, of Hermann Weyl's definitive work on Riemann surfaces [W_2]. At that time, and for some time thereafter, all surfaces under consideration are assumed to be orientable. Schiffer and Spencer return to Klein's concept in their 1954 monograph [SS] and thus give the non-orientable case its first modern treatment. On the algebraic side, L. Berzolari [B] gave an account on plain algebraic curves, at about the turn of the century, which did treat curves over the reals. However his work was severely handicapped by the lack of valuation theory, which was not then used well enough to treat the real case with ease. By that time algebraic geometry had generalized itself to a very great extent and had left real-algebraic curves far behind.

The fact that any real-algebraic function field E, whose constant field is R, is just the fixed field of $F = E(i)$ under an R-linear automorphism σ of order two, and thus that the Klein surface $\mathfrak{X}$ of E is just the orbit space of the Riemann surface $\mathfrak{W}$ of F under σ^* -- is obvious, and was well known. Our thesis has been that it is better to work with E on $\mathfrak{X}$ rather than to have to work with E imbedded in $E(i)$ on $\mathfrak{B}$. Hopefully this monograph and successive work in this direction will give weight to our point of view. Already application has been made by the first author to the theory of real-Banach algebras [A_5]. These preliminary remarks were elaborated by the first author and L. Andrew Campbell in [AC] to prove an extension of the Arens-Royden Theorem to real Banach algebras. Another offshoot of this theory, that of analytic and harmonic obstruction on non-compact Klein surfaces, has been considered by the first author [A_6]. In [A_7] he is making a study of algebraic analytic obstruction theory on compact Klein surfaces and the way it reflects the topological obstruction on the underlying surface. The second author has dealt with the foundations of sheaf theory on Klein surfaces [G_1]; and he, together with Walter Read have investigated the question of the existence of positive differentials on elliptic or hyperelliptic Klein surfaces with boundary [GR]. Further work is contemplated.

Many mathematicians aided us in our bibliographic researches. Professors Ahlfors and Bers and Sir Edward Collingwood were particularly helpful. We would also like to thank our colleague Professor Sanford Segal (of Rochester) for bringing the paper by Berzolari to our attention.

The first author would like to thank the National Science Foundation for its support, under NSF Grant GP 9214, and the University of California at San Diego for the hospitality afforded him during the fall of 1969. Many thanks also to Patricia Pattison and Patricia Cristantello for the care they lavished on typing this manuscript.

Rochester, New York
Würzburg, Bavaria

Austin, Texas

June, 1971

TABLE OF CONTENTS

CHAPTER 1. KLEIN SURFACES

§1. Analytic Preliminaries

Let A be a nonempty open subset of R^2 and let f be C^2-map of A into R^2. Recall that $\partial/\partial z \equiv (\partial/\partial x - i\,\partial/\partial y)/2$ and that $\partial/\partial\bar{z} \equiv (\partial/\partial x + i\,\partial/\partial y)/2$. We shall ususally identify R^2 with the complex plane C. However, in referring to f as a C^2-map we shall mean that its real and imaginary parts are real-valued C^2-maps of $A \subset R^2$. f is <u>analytic</u> on A if $\partial f/\partial\bar{z} = 0$ on A and <u>antianalytic</u> on A if $\partial f/\partial z = 0$ on A. It will also be convenient to regard the empty map as being analytic and antianalytic. Since $\partial/\partial\bar{z}$ and $\partial/\partial z$ are C-linear derivations the set of all analytic functions, and the set of all antianalytic functions are C-algebras, under pointwise operations. f will be called <u>dianalytic</u> if on each connected component V of A, f is analytic or antianalytic. If f is both analytic and antianalytic on V, then f is easily seen to be constant on V. Assume that f is a nonconstant dianalytic function on V. Then let $\tau_V(f) \equiv 0$ if f is analytic and let $\tau_V(f) \equiv 1$ if f is antianalytic on V.

Lemma 1.1.1. $\overline{\dfrac{\partial f}{\partial z}} = \dfrac{\partial \bar{f}}{\partial \bar{z}}$; thus f (resp. $\bar{f}$) is analytic if and only if $\bar{f}$ (resp. f) is antianalytic.

Proof. Let u and v be the real and imaginary parts of f, respectively. $\partial f/\partial z = (\partial u/\partial x + i\,\partial v/\partial x - i\,\partial u/\partial y + \partial v/\partial y)/2$, and $\partial\bar{f}/\partial\bar{z} = (\partial u/\partial x - i\,\partial v/\partial x + i\,\partial u/\partial y + \partial v/\partial y)/2$; proving the lemma.

Let $f(A)$ be contained in the open set B, g be a C^2-map of B into $\mathbb{R}^2$, $h \equiv g(f)$; and let $w \equiv f(z)$, the complex variable w being written as $u + iv$, u and v being real.

Lemma 1.1.2. $\partial h/\partial\bar{z} = (\partial g/\partial w)(\partial f/\partial\bar{z}) + (\partial g/\partial\bar{w})(\partial\bar{f}/\partial\bar{z})$ and $\partial h/\partial z = (\partial g/\partial w)(\partial f/\partial z) + (\partial g/\partial\bar{w})(\partial\bar{f}/\partial z)$.

Proof. $\partial h/\partial\bar{z} = \partial g(f)/\partial\bar{z} = (\partial g/\partial u)(\partial u/\partial\bar{z}) + (\partial g/\partial v)(\partial v/\partial\bar{z})$

$$= 2^{-1}(\partial g/\partial u - i\,\partial g/\partial v)\partial f/\partial\bar{z} + 2^{-1}(\partial g/\partial u + i\,\partial g/\partial v)\partial\bar{f}/\partial\bar{z}$$

$$= (\partial g/\partial w)(\partial f/\partial\bar{z}) + (\partial g/\partial\bar{w})(\partial\bar{f}/\partial\bar{z}),$$

proving the first assertion. Concerning the second equality, note that, by (1.1.1) $\frac{\partial h}{\partial z} = \overline{\overline{\frac{\partial h}{\partial z}}} = \overline{\frac{\partial\bar{h}}{\partial\bar{z}}} = \overline{\frac{\partial\bar{g}}{\partial w}}\,\overline{\frac{\partial f}{\partial\bar{z}}} + \overline{\frac{\partial\bar{g}}{\partial\bar{w}}}\,\overline{\frac{\partial\bar{f}}{\partial\bar{z}}} = \frac{\partial g}{\partial\bar{w}}\frac{\partial\bar{f}}{\partial z} + \frac{\partial g}{\partial w}\frac{\partial f}{\partial z}$, proving the lemma.

Proposition 1.1.3. Let f and g be dianalytic; then $h \equiv g(f)$ is dianalytic. Assume now that neither f nor g is constant and that A and B are connected. Then $\tau_A(h) = \tau_B(g) + \tau_A(f) \pmod 2$.

Proof. Without loss of generality we may assume that A and B are connected. If either f or g is constant then so is h, and hence is dianalytic. Assume neither f nor g is constant. If $\tau_A(f) = 0 = \tau_B(g)$, then we can appeal to the first equation in (1.1.2) to see that $\tau_A(h) = 0$. If $\tau_A(f) = 1 = \tau_B(g)$, then $\partial(g)/\partial w = 0$ and, by (1.1.1), $\partial\bar{f}/\partial\bar{z} = \overline{\partial f/\partial z} = \bar{0} = 0$; hence $\tau_A(h) = 0$. A similar proof works in the other cases.

Recall that the Laplacean operator, Δ, is $\partial^2/\partial x^2 + \partial^2/\partial y^2$, and that a C^2-function f on A is harmonic if $\Delta f = 0$. Since $4(\partial/\partial z)(\partial/\partial\bar{z}) = \Delta = 4(\partial/\partial\bar{z})(\partial/\partial z)$, every dianalytic function is harmonic. It is well known that if f is antianalytic then $\bar{f}$ is analytic, and u and $-v$ are harmonic. Thus f dianalytic implies u and v are harmonic. If f is analytic then u and v, in addition to being harmonic, are harmonic conjugates: i.e., $\partial u/\partial x = \partial v/\partial y$ and $\partial u/\partial y = -\partial v/\partial x$. Further these conditions are sufficient to insure that f is analytic. Let f now be antianalytic on A; then we see that $\partial u/\partial x = -\partial v/\partial y$ and $\partial u/\partial y = \partial v/\partial x$: i.e., that u and v are <u>harmonic</u> <u>anticonjugates</u>. Clearly if u and v are harmonic anticonjugates then f is antianalytic.

Theorem 1.1.4. If g is harmonic and f dianalytic, then $h \equiv g(f)$ is harmonic. Conversely, if for all harmonic g, $h \equiv g(f)$, is harmonic, then f is dianalytic.

Proof. One can easily check that

$$(*) \quad \Delta h = (\partial^2 g/\partial u^2)|\text{grad } u|^2 + (\partial^2 g/\partial v^2)|\text{grad } v|^2 + 2(\partial^2 g/\partial u\partial v)(\text{grad } u)\cdot(\text{grad } v) + (\partial g/\partial u)\Delta u + (\partial g/\partial v)\Delta v.$$

Assume that f is dianalytic. Then $|\text{grad } u| = |\text{grad } v|$, $(\text{grad } u)\cdot(\text{grad } v) = \Delta u = \Delta v = 0$; thus $\Delta h = (\Delta g)|\text{grad } u|^2$. Assume now that g is harmonic. Then $\Delta h = 0$, proving the first statement. To prove the converse, first use $g = u$ and $g = v$ to conclude that $\Delta u = \Delta v = 0$. Then, since h is harmonic, (*) becomes

$$(**) \quad 0 = \Delta h = \frac{\partial^2 g}{\partial u^2}|\text{grad } u|^2 + \frac{\partial^2 g}{\partial v^2}|\text{grad } v|^2 + 2\left(\frac{\partial^2 g}{\partial u\partial v}\right)(\text{grad } u)\cdot(\text{grad } v).$$

Now if $g(u,v) = u^2 - v^2$, (**) yields $|\text{grad } u|^2 = |\text{grad } v|^2$, while if $g(u,v) = uv$, then we obtain $(\text{grad } u)\cdot(\text{grad } v) = 0$. Hence grad u and grad v are always equal in length and perpendicular. Hence for each $\alpha \in A$, we have $(\partial f/\partial z)(\alpha) = 0$, or $(\partial f/\partial \bar{z})(\alpha) = 0$. The following lemma completes the proof.

Lemma 1.1.5. Let f be a complex-valued function defined on a connected, open subset A of C, whose real and imaginary parts, u and v are C^2-maps. Assume that for all $\alpha \in A$, $(\partial f/\partial \bar{z})(\alpha) = 0$ or $(\partial f/\partial z)(\alpha) = 0$; then u and v are harmonic, and f is analytic on all of A or antianalytic on all of A.

Proof. Since f is a C^2-map, Δf is well defined, and equals $4 \frac{\partial}{\partial z} \frac{\partial f}{\partial \bar{z}} = 4 \frac{\partial}{\partial \bar{z}} \frac{\partial f}{\partial z}$, which by hypothesis is zero. Thus f, and hence u and v are harmonic. Let $U = \{\alpha \in A : \partial f/\partial z(\alpha) \neq 0\}$. Then U is open in A, and f is analytic on U. If $U = \bar{U}$, the closure of U in A, then either $U = A$ or $U = \emptyset$, and we are through. Let $\lambda \in \bar{U} - U$, and let V be a simply connected neighborhood of λ in A. Then u has a harmonic conjugate w on V and we may assume that $w(\lambda) = v(\lambda)$. Since f is analytic on U, $w - v$ must be constant on each component of $U \cap V$, and hence $w - v$, being harmonic, is constant on V: thus $w = v$ on V. Thus f is analytic on all of V, and in particular f is analytic at λ. Since $\lambda \notin U$, $\partial f/\partial z(\lambda) = 0$. But this is merely $f'(\lambda)$. Since f is analytic at λ, the set of all $\alpha \in V$ for which $f'(\alpha) = 0$ is discrete; thus $\bar{U} - U$ is discrete. Hence $\bar{U}$ is open and closed in A. Since A is connected either $\bar{U} = \emptyset$, or $\bar{U} = A$. This concludes the proof.

Note that f is said to be dianalytic if f is analytic or antianalytic on each component of A; thus the definition is not, a priori, a local one. But now, we have

Corollary 1.1.6. Let f be a locally dianalytic function on A; then f is a dianalytic function on A.

§2 Dianalytic Atlases and Klein Surfaces

Let $C^+ \equiv \{a + bi : a,b \in R \text{ and } b \geq 0\}$ be referred to as the upper half plane. Let A and B be nonempty open sets in C^+ and let f be a continuous map of A into B. f is said to be analytic on A (resp. antianalytic on A) if it extends to an analytic (resp. antianalytic) function on some neighborhood of A in C into C. If f is analytic or antianalytic on each component of A, then f is said to be dianalytic on A.

Proposition 1.2.1 (The Schwartz Reflection Principal). Let A and B be open sets in C^+ and let $f : A \longrightarrow B$ be continuous, dianalytic on the interior of A, and satisfy $f(A \cap R) \subset B \cap R$. Then f is dianalytic on A.

Proof: Let $\overline{A} \equiv \{\overline{z} \mid z \in A\}$. Then f extends to a dianalytic map $\overline{f} : A \cup \overline{A} \longrightarrow B \cup \overline{B}$ by setting $\overline{f}(\overline{z}) = \overline{f(z)}$; see e.g., [SZ].

Let X be a non-empty connected Hausdorff space. If there exists a family $\underset{\sim}{U} = (U_j, z_j)_{j \in J}$, where $(U_j)_{j \in J}$ is an open cover of X, and z_j is a homeomorphism of U_j onto an open set in either C or C^+, then X is called a two-manifold with boundary. The boundary ∂X of X consists of those points $x \in X$ such that $x \in U_j$ with $z_j(U_j)$ open in C^+, but not open in C, and $z_j(x) \in R$. Then $\underset{\sim}{U}$ is called an atlas of X. The maps $z_j z_k^{-1} : z_k(U_j \cap U_k) \longrightarrow z_j(U_j \cap U_k)$ are surjective homeomorphisms, that are called the transition functions of $\underset{\sim}{U}$. $\underset{\sim}{U}$ will be called a dianalytic atlas (resp. analytic atlas) if all of its transition functions are dianalytic (resp. analytic). Each pair (U_j, z_j) is called a chart of $\underset{\sim}{U}$. Clearly if $\underset{\sim}{U}$ is analytic, then it is dianalytic.

Let $\underset{\sim}{U} = (U_j, z_j)_{j\in J}$ and $\underset{\sim}{V} = (V_k, y_k)_{k\in K}$ be dianalytic (resp. analytic) atlases of the two-manifold X. $\underset{\sim}{U}$ and $\underset{\sim}{V}$ are dianalytically (resp. analytically) equivalent if $\underset{\sim}{U} \cup \underset{\sim}{V}$ is dianalytic (resp. analytic); in this case write $\underset{\sim}{U} \sim \underset{\sim}{V}$.

Example 1.2.1. Let $X = C = U_1 = U_2$ and let $z_1(z) = z$, $z_2(z) = \bar{z}$. Let $\underset{\sim}{U}_1 = \{(U_1, z_1)\}$, $\underset{\sim}{U}_2 = \{(U_2, z_2)\}$, and $\underset{\sim}{U}_3 = \{(U_1, z_1), (U_2, z_2)\}$. Then $\underset{\sim}{U}_1$ and $\underset{\sim}{U}_2$ are analytic atlases, whereas $\underset{\sim}{U}_3$ is a dianalytic atlas which is not analytic. $\underset{\sim}{U}_1$ and $\underset{\sim}{U}_2$ are dianalytically equivalent but not analytically equivalent.

Lemma 1.2.2. The relation of dianalytic (resp. analytic) equivalence is an equivalence relation.

An equivalence class $\mathfrak{X}$ of dianalytic (resp. analytic) atlases of X will be called a dianalytic (resp. analytic) structure on X. A pair consisting of a two-manifold with boundary X and dianalytic structure $\mathfrak{X}$ on X will be called a Klein surface, and will often be denoted by $\mathfrak{X}$. By a Riemann surface we shall mean a pair consisting of a two-manifold X (without boundary) and an analytic structure $\mathfrak{X}$ on X. We shall further need the terms Klein surface without boundary and Riemann surface with boundary, whose meanings should be clear.

Bibliographic note. The problem of understanding scripture of a distant era, written in a foreign tongue, is not merely one of translation, but one of recasting antique ideas into modern form; similarly, for example, it is hard to find an acceptable modern definition of the notion of a Riemann surface in Riemann's work. However, it is certainly correct to attribute this idea to Riemann. The notion of a Klein surface seems attributable to Klein by dint of his remarks in 1882 on the closing pages of his Über Riemanns Theorie der algebraischen Functionen und ihrer Integrale [K], even though one does not find a definition (acceptable to a modern reader) of a Klein surface there. While Riemann surfaces have been extensively investigated during the last century or so, the work on Klein surfaces, which are not Riemann surfaces -- since their introduction by Klein in 1882 -- seems to have been at best sporadic, until the appearance of Schiffer and Spencer's extensive work on Functionals of Finite Riemann Surfaces [S S], in 1954. To avoid confusion we have gone back to an old tradition [B], that of calling the surfaces Klein introduced "Klein-Riemannschen Flache," by calling them Klein surfaces, rather than using Schiffer and Spencer's terminology: i.e., calling all such surfaces, Riemann surfaces.

Example 1.2.2. Möbius strips as Klein surfaces. Consider the "half-open" rectangle in C defined by $0 < \mathrm{Re}\, z < 1$, $0 < \mathrm{Im}\, z < a$, where $a > 0$. We form a Möbius strip X by identifying a small strip on the left with one on the right as indicated. Then cover X by open sets U_i, $1 \leq i \leq 6$, as indicated. Now set $z_i(z) = z$ for 1, 2, 3, and $z_i(z) = ai - z$ for $i = 4, 5, 6$. It is easily checked that this gives a dianalytic structure on X, with the transition functions analytic except for $z_j z_k^{-1}$, where $(k,j), (j,k) \in \{1,4\} \times \{3,6\}$.

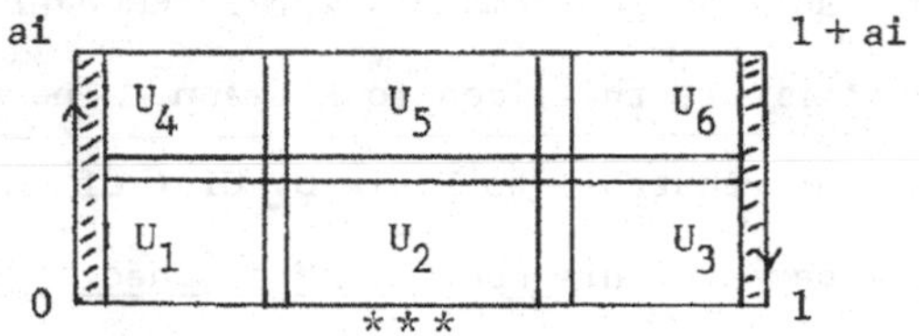

Let g be a continuous real or complex-valued function on X. g will be said to be a C^n - function relative to $\underset{\sim}{U}$. $(0 \leq n \leq \infty)$, if for all $j \in J$, $(g \mid U_j^0) \circ z_j^{-1}$ is a C^n - function, where U_j^0 denotes $U_j - \partial X$. Let $\underset{\sim}{V} = (V_k, y_k)_{k \in K} \in \mathfrak{X}$. Since the transition functions $z_j y_k^{-1}$ are C^∞- maps, this definition is independent of the choice $\underset{\sim}{U}$, and thus gives rise to the notion of a C^n - function on $\mathfrak{X}$.

Let g be a C^2 - function on $\mathfrak{X}$. g will be called harmonic relative to $\underset{\sim}{U}$ if each $(g \mid U_j^0) \circ z_j^{-1}$ is harmonic. Since the transition functions $z_j y_k^{-1}$ are dianalytic we may apply Theorem 1.1.4 and conclude that g is harmonic relative to $\underset{\sim}{U}$ if and only if it is harmonic relative to $\underset{\sim}{V}$. Hence, dependence on $\underset{\sim}{U}$ is inessential, and as a result if g is harmonic relative to some $\underset{\sim}{U} \in \mathfrak{X}$, we say that g is harmonic on $\mathfrak{X}$.

Theorem 1.2.3. The notion of a harmonic function, as being a solution of Laplace's equation, makes sense on a Klein surface. Conversely, a Klein surface is the most general two-manifold in which this notion of harmonic functions makes sense.

Proof. The first statement has just been proved. As to the second, it is a direct corollary of the second half of Theorem 1.1.4; proving the theorem.

Let $\mathfrak{Y}$ be a Klein surface and let $\underset{\sim}{W} \equiv (W_m, x_m)_{m\in M} \in \mathfrak{Y}$. Let g be a continuous map of X into Y. g will be called a C^n-map (resp. a harmonic map) if for all $(j,m) \in J \times M$,

$$x_m \cdot (g \mid U_j^0 \cap g^{-1} W_m^0) \cdot z_j^{-1}$$

is a C^n-map (resp. harmonic). As noted above this definition is independent of the choice of $\underset{\sim}{U} \in \mathfrak{X}$ and $\underset{\sim}{W} \in \mathfrak{Y}$.

Having met with such success in defining harmonic functions and harmonic maps, the reader might now expect a discussion at this point of analytic functions and maps on Klein surfaces, for such notions exist in the theory of Riemann surfaces. Let g be a continuous complex-valued function on X. For the moment, g will be called analytic relative to $\underset{\sim}{U}$ if each $(g \mid U_j) \cdot z_j^{-1}$ is analytic.

Example 1.2.3. Let $U_1 = C = U_2$, let $z_1 = z$, and $z_2 = \bar{z}$. Thus $\underset{\sim}{U} \equiv (U_j, z_j)_{j\in\{1,2\}}$ is a dianalytic coordinate covering of C, thus $\mathfrak{C}$ is a Klein surface. Let $g = z$, note that $g \cdot z^{-1}$ is analytic, whereas $g \cdot z_2^{-1}$ is antianalytic. g, and in fact nonconstant convergent power series in g, are not analytic on $\mathfrak{C}$. Clearly this could be remedied by letting $\underset{\sim}{U}_j \equiv (U_j, z_j)$, $j = 1$, and 2. Then g is analytic relative to $\underset{\sim}{U}_1$ but not to $\underset{\sim}{U}_2$.

While the last example may be viewed as the result of a capricious definition, the following is more serious.

Example 1.2.4. Let X be a connected nonorientable Klein surface, and let g be a meromorphic function on X, relative to $\underset{\sim}{U} \in \mathfrak{X}$, in the sense of the definition above; then g is constant. Since X is nonorientable there exist open connected sets $(U_i')_{i=1,\ldots,n}$; such that for each i, $U_i' \subset U_i$ and only one j exists other than i such that $U_i' \cap U_j' \neq \emptyset$, this set being connected such that $1 \leq i < n$ implies $j = i + 1$, and $i = n$ implies $j = 1$; for which $\tau_{U_n' \cap U_1'}(z_n z_1^{-1}) + \Sigma_{j=1}^{n-1} \tau_{U_j' \cap U_{j+1}'}(z_j z_{j+1}^{-1}) \equiv 1 \pmod{2}$. Thus there exist distinct i and j such that

$$U_i' \cap U_j' \neq \emptyset \quad \text{and} \quad \tau_{U_i' \cap U_j'}(z_i z_j^{-1}) = 1.$$

Since g is meromorphic in the sense above, $(g \mid U_i') \cdot z_i^{-1}$ and $(g \mid U_j') \cdot z_j^{-1}$ are meromorphic. Hence $(g \mid U_i' \cap U_j') \cdot z_i^{-1} \cdot z_i z_j^{-1}$ is $(g \mid U_i' \cap U_j') \cdot z_j^{-1}$ and as a consequence is antimeromorphic and meromorphic simultaneously: i.e., $g \mid U_i' \cap U_j'$ is constant. Since X is connected we may employ analytic continuation to prove that g is constant on all of X.

We conclude that the notion of an analytic function as given above is not appropriate for use on a Klein surface which is not a Riemann surface.

Theorem 1.2.4. Let $\mathfrak{X}$ be an orientable Klein surface. There are two analytic structures $\mathfrak{X}_1$ and $\mathfrak{X}_2$ on X which are dianalytically equivalent to $\mathfrak{X}$.

Proof. Let $\underset{\sim}{U} = (U_j, z_j)_{j \in J} \in \mathfrak{X}$. We may assume that all U_j are connected and that if $(U,z) \in \underset{\sim}{U}$, then so is $(U,-\bar{z})$. Since X is orientable, each transition function of $\underset{\sim}{U}$ is either analytic or antianalytic. Further, if $z_j z_k^{-1}$ and $z_i z_j^{-1}$ are both analytic, then so is $z_k z_i^{-1}$. Let $\underset{\sim}{V}$ be a maximal subfamily of $\underset{\sim}{U}$ such that all transition functions in $\underset{\sim}{V}$ are analytic, and let $(U,z) \in \underset{\sim}{U}$ with $(U,z) \notin \underset{\sim}{V}$. Then there exists $(U_j,z_j) \in \underset{\sim}{V}$ with $z_j z^{-1}$ antianalytic and hence $z_i z^{-1}$ is antianalytic for all $(U_i,z_i) \in \underset{\sim}{V}$. Hence $(U,-\overline{z}) \in \underset{\sim}{V}$, and $\underset{\sim}{V}$ is an analytic atlas on X. It is easily seen that any analytic atlas in $\mathfrak{X}$ is analytically equivalent either to $\underset{\sim}{V}$ or to $\overline{\underset{\sim}{V}} = \{(U,-\overline{z}) \mid (U,z) \in \underset{\sim}{V}\}$.

Thus if $\mathfrak{X}$ is an orientable Klein surface there are two induced analytic structures $\mathfrak{X}_1$ and $\mathfrak{X}_2$ on X. Choosing between $\mathfrak{X}_1$ and $\mathfrak{X}_2$ is equivalent to choosing an orientation for X.

§3 Meromorphic functions

In order to define a meromorphic function on a Klein surface X we first need to extend complex conjugation, $a + bi \to a - bi$, to the Riemann sphere by sending ∞ to ∞; thus the real line and ∞ are the fixed points of this map. Next, on non-orientable Klein surfaces, non-constant meromorphic functions do not exist in the usual sense, hence we must resort to some subterfuge in order to make an interesting definition. The following may be found in Schiffer and Spencer [SS]. Actually this definition is natural and is forced algebraically, as we will see in Chapter 2.

A meromorphic function $f_{\underset{\sim}{U}}$ on X relative to $\underset{\sim}{U} \equiv (U_j, z_j)_{j\in J} \in \mathfrak{X}$ is a family $(f_j)_{j\in J}$ of maps of the U_j's into Σ, such that each $f_j z_j^{-1}$ is meromorphic on $z_j(U_j)$ and $f_j(\partial X \cap U_j) \subset R \cup \{\infty\}$, subject to the following compatibility condition: given a connected set V of $U_j \cap U_k$, if $z_j z_k^{-1} \mid z_k(V)$ is analytic (resp. antianalytic) then $f_j \mid V = f_k \mid V$ (resp. $f_j \mid V = \overline{f}_k \mid V$).

Example 1.3.1. Assume that X is non-orientable. Let $a, b \in R$, $b \neq 0$ and let $f_j = \alpha = a + bi$ for all $j \in J$. Then $f_{\underset{\sim}{U}} \equiv (f_j)_{j\in J}$ is not a meromorphic function on X relative to $\underset{\sim}{U}$, for there is some non-empty connected subset V of $U_j \cap U_k$ such that $z_j z_k^{-1}$ is antianalytic on $z_k(V)$; thus α must be $\overline{\alpha}$, which is absurd.

Of course, if we let $f_j \equiv a$ for all $j \in J$, $f_{\underset{\sim}{U}}$ is indeed a meromorphic function on $\mathfrak{X}$ relative to $\underset{\sim}{U}$. The existence of a non-constant meromorphic function on X relative to $\underset{\sim}{U}$ is of course a deep question which we will not settle now. It will be settled in the affirmative in §6.

It is clear that the set of all meromorphic functions on $\mathfrak{X}$ relative to $\mathbf{U}$ form a field extension of R, under pointwise operations.

In order to free the notion of a meromorphic function on $\mathfrak{X}$ from a particular coordinate covering we introduce the following equivalence relation. Let $\mathbf{V} \equiv (V_k, w_k)_{k \in K}$ be in $\mathfrak{X}$ and let $\mathbf{g_V} \equiv (g_k)_{k \in K}$ be a meromorphic function on $\mathfrak{X}$ relative to $\mathbf{V}$. We will write $\mathbf{f_U} \sim \mathbf{g_V}$ and say that $\mathbf{f_U}$ is <u>equivalent</u> to $\mathbf{g_V}$ if $\mathbf{f_U} \cup \mathbf{g_V}$ is a meromorphic relative to $\mathbf{U} \cup \mathbf{V}$. This is easily seen to be an equivalence relation.

Lemma 1.3.1. If $\mathbf{f_U}$ is a meromorphic function on $\mathfrak{X}$ relative to $\mathbf{U}$ and (V,w) is a dianalytic chart on $\mathfrak{X}$, then there is a $\mathbf{V}$ and a unique function $\mathbf{f_V}$ on $\mathbf{V}$ such that $\mathbf{f_U} \cup \mathbf{f_V}$ is a meromorphic function relative to $\mathbf{U} \cup (V,w)$.

Proof: On $V \cap U_j$ set $f_V = f_j$ when wz_j^{-1} is analytic and $f_V = \overline{f}_j$ when wz_j^{-1} is antianalytic.

Corollary. If $\mathbf{U}, \mathbf{V} \in \mathfrak{X}$, then the fields of meromorphic functions on X relative to $\mathbf{U}$ and $\mathbf{V}$ are canonically isomorphic.

Thus the equivalence classes of meromorphic functions with respect to different $\mathbf{U} \in \mathfrak{X}$ form a field, known as the field of meromorphic functions on $\mathfrak{X}$ and denoted by $E(\mathfrak{X})$.

Theorem 1.3.2. $E(\mathfrak{X})$ contains a copy of C if and only if X is orientable and $\partial X = \emptyset$.

Proof. In Example (1.3.1) we saw that X non-orientable implies that $i \notin E(\mathfrak{X})$; thus $C \subset E(\mathfrak{X})$ implies X orientable. Now assume X to be orientable. By (1.2.4), X admits an analytic structure $\mathfrak{X}_1 \subset \mathfrak{X}$. Given $\underline{U} = (U_j, z_j)_{j \in J} \in \mathfrak{X}_1$, let $f_j = i$ for all $j \in J$ and note that $(f_j)_{j \in J}$ is a meromorphic function $f_{\underline{U}}$ on X relative to $\underline{U}$. Since $f^2 = -1$ and $R \subset E(\mathfrak{X})$, we see that $C \subset E(\mathfrak{X})$, proving the theorem.

Let $f_{\underline{U}} \equiv (f_j)_{j \in J}$ be a meromorphic function on X relative to $\underline{U} \in \mathfrak{X}$. It will be called analytic (or holomorphic) if $f_j(U_j) \subset C$ for all $j \in J$. Clearly this is independent of the choice of $\underline{U}$. Let $\mathcal{H}(\mathfrak{X})$ denote the set of all $\underline{f} \in E(\mathfrak{X})$ that are holomorphic.

Corollary 1.3.3. $\mathcal{H}(\mathfrak{X})$, an R - algebra, is an integral domain. Finally, assume that $\partial X = \emptyset$; then $\mathcal{H}(\mathfrak{X})$ contains a copy of C if and only if X is orientable.

Let $\underline{f} \in E(\mathfrak{X})$, $\underline{f} \neq 0$. For $x \in X$ choose a dianalytic chart (U,z) with $x \in U$. If $x \in \partial X$ extend $f_U z^{-1}$ across the real axis by Schwarz reflection. Let $\nu_x(\underline{f})$ be the order of $f_U z^{-1}$ (extended if necessary) at $z(x)$. It is easily checked that this is well defined. We say that x is a pole of $\underline{f}$ if $\nu_x(\underline{f}) < 0$. Clearly the poles of $\underline{f}$ form a discrete subset of X.

As noted before, meromorphic functions are not maps, but equivalence classes of compatible families of maps. We can, however, associated with $\underline{f} \in E(\mathfrak{X})$ a map f of X into $\mathrm{cl}\, C^+$ which has its uses, $\mathrm{cl}\, C^+$ being the closure of C^+ in Σ. In order to do this let $f_{\underline{U}} \in \underline{f}$. Let $P_{\underline{f}} \equiv \{x \in X : \underline{f}$ has a pole at $x\}$. Away from the poles of $\underline{f}$, f_j may be written in a unique fashion as $a_j + b_j i$, where a_j and b_j are real-valued functions. Further, $a_j z_j^{-1}$ and $b_j z_j^{-1}$ are harmonic conjugates. Since $(f_j)_{j\in J}$ is a meromorphic function on X relative to $\underset{\sim}{U}$, $a_j \mid U_j \cap U_k = a_k \mid U_j \cap U_k$. As a consequence, we may patch the maps $(a_j)_{j\in J}$ together to form a map a of X which is harmonic on $X - P_{\underline{f}}$.

The family $(b_j)_{j\in J}$ does not necessarily patch together to form a global map. However, $|b_j| \Big| U_j \cap U_k = |b_k| \Big| U_j \cap U_k$; hence $(|b_j|)_{j\in J}$ does patch together to form a continuous map c of $X - P_{\underline{f}}$ into $R^+ \equiv \{r \in R : r \geq 0\}$. Let $f \equiv a + ci$ on $X - P_{\underline{f}}$, and let $f(x) = \infty$ on $P_{\underline{f}}$.

Lemma 1.3.4. f is a well-defined continuous map of X into cl C^+, which is independent of the choice of $\underset{\sim}{U}$.

Proof. Let $\underset{\sim}{V} \equiv (V_k, w_k)_{k\in K} \in \mathfrak{X}$ such that $f_{\underset{\sim}{V}} \in \underline{f}$. Let $x \in V_k$ and note that $f_j(x) = f_k(x)$ or $\overline{f_k(x)}$, proving the lemma.

Example 1.3.2. i and -i are elements in $E(\mathfrak{C})$ and yet the maps of each function is the map, $\lambda \to i$ for all $\lambda \in C$.

Example 1.3.3. Consider the function $\underline{f}$ in $E(\Sigma)$ commonly referred to as z. The map f associated with $\underline{f}$ has the following description: given $a + bi \in C$, $a,b \in R$, $f(a + bi) = a + |b|i$; further $f(\infty) = \infty$. f, which will be referred to as the <u>folding map</u> φ_{C^+} or φ, is the identity on cl C^+ and folds the lower half plane onto the upper half plane. Clearly, $R \cup \{\infty\}$ is the set of points where folding occurs, and this is exactly the set of points at which harmonicity for f fails.

Let $\underline{f} \in E(\mathfrak{X})$ be nonconstant, and let f be its map. Let $\Gamma_{\underline{f}} \equiv \{x \in X: f(x) \in R \cup \{\infty\}\}$, and let this be called the real locus of $\underline{f}$. Note that on $X \setminus \Gamma_{\underline{f}}$ the absolute value of the imaginary part of $\underline{f}$ is harmonic and real-valued.

Theorem 1.3.5. Let $\underline{f}$ be a non-constant element in $E(\mathfrak{X})$, and let $\Gamma_{\underline{f}}$ be the real locus of $\underline{f}$. Each component of $X \setminus \Gamma_{\underline{f}}$ is orientable.

Proof. The restriction of f to $X \setminus \Gamma_{\underline{f}}$ maps it as a ramified covering to the interior of $\mathrm{cl}\ C^+$. Hence the orientation of $\mathrm{cl}\ C^+$ lifts, via f, to each component of $X \setminus \Gamma_{\underline{f}}$.

Clearly $\partial X \subset \Gamma_{\underline{f}}$ for all non-constant $\underline{f} \in E(\mathfrak{X})$. If X is non-orientable, then so is $X - \partial X$; thus by the last Theorem, it is impossible for $\partial X = \Gamma_{\underline{f}}$ in this case.

Theorem 1.3.6. (Ahlfors [A_1]). Let $\mathfrak{X}$ be compact, connected, orientable Klein surface with non-void boundary. There exists $\underline{f} \in E(\mathfrak{X})$ such that $\partial X = \Gamma_{\underline{f}}$.

Corollary 1.3.7. Let $\mathfrak{X}$ be a compact, connected Klein surface with a non-void boundary. X is orientable if and only if there exists $\underline{f} \in E(\mathfrak{X})$ such that $\partial X = \Gamma_{\underline{f}}$.

§4 Morphisms

We now wish to define a morphism $f : \mathfrak{X} \longrightarrow \mathfrak{Y}$ of Klein surfaces. This differs from the corresponding concept for Riemann surfaces principally in that X may "fold" along ∂Y. Recall that the folding map $\varphi : C \longrightarrow C^+$ is given by $\varphi(a + bi) = a + |b|i$. For technical reasons we need the notion of a positive chart (U,z), it being a chart such that $z(U) \subset C^+$.

Definition. A morphism $f : \mathfrak{X} \longrightarrow \mathfrak{Y}$ of Klein surfaces is a continuous map f of X into Y, with $f(\partial X) \subset \partial Y$, such that for all $x \in X$ there exist dianalytic charts (U,z) and (V,w) about x and $f(x)$ respectively, and an analytic function F on $z(U)$ such that the following diagram commutes:

(1.4.1)
$$\begin{array}{ccccc} U & & \xrightarrow{\quad f \quad} & & V \\ \Big\downarrow z & & & & \Big\downarrow w \\ C^+ & \xrightarrow{F} & C & \xrightarrow{\varphi} & C^+ \end{array}$$

(Note that the chart (V,w) must be positive.)

Note that since $f(\partial X) \subset \partial Y$, F must be real-valued on $R \cap z(\partial X \cap U)$; hence F can be extended to an analytic function $\hat{F}$ on $z(U) \cup \overline{z(U)}$, by setting $\hat{F}(z) = \overline{F(\bar{z})}$, for all $z \in \overline{z(U)}$.

Lemma 1.4.1. $\varphi F \varphi = \varphi \hat{F}$.

Proof. Since $\hat{F}$ extends F, the equation holds on $z(U)$. Let $\lambda \in z(U)$, and let $\mu = \bar{\lambda}$. $\varphi F \varphi(\mu) = \varphi F(\lambda)$. Whereas, $\varphi \hat{F}(\mu) = \varphi \overline{F}(\lambda) = \varphi F(\lambda)$, proving the lemma.

Remark. In case $\partial Y = \emptyset$ we may replace diagram (1.4.1) with the following simpler one:

(1.4.1a)

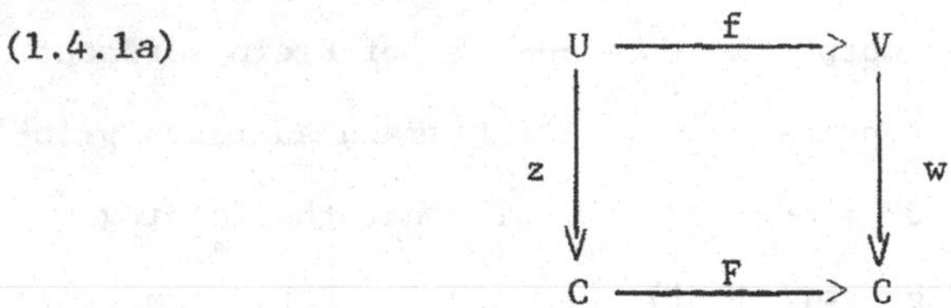

Rather than dealing with cases, we shall usually use (1.4.1).

Example 1.4.1. The folding map $\varphi : C \longrightarrow C^+$ is a morphism. Let n be a positive integer, let $z(\lambda) = \lambda$, $F(\lambda) = \lambda^n$, and $w(\lambda) = \lambda$ for all $\lambda \in C^+$. Then the corresponding f is a morphism of C^+ onto C^+.

Lemma 1.4.2. Let $f : \mathfrak{X} \longrightarrow \mathfrak{Y}$ be a non-constant morphism, and let (U,z), (V,w) be analytic charts with $f(U) \subset V$, and (V,w) positive. Then there exists a unique analytic function F on $z(U)$ such that $wf = \varphi Fz$.

Proof. Suppose that U is covered by open sets U_j, and that on $z(U_j)$ we can find an analytic function F_j with $wf = \varphi F_j z$ on U_j. Since F_j is non-constant and analytic on $z(U_j \cap U_k)$ either $F_j = F_k$ or $F_j = \overline{F}_k$. We must have $F_j = F_k$, and so the F_j patch together to give the desired function F on $z(U)$.

Hence we may assume that there exist functions z_1, w_1, F_1 on U, V, $z_1(U)$, respectively, so that (U,z_1), (V,w_1) are dianalytic charts, F_1 is analytic, and $w_1 f = F_1 z_1$. Since (V,w) and (V,w_1) are positive, we may define $\widehat{w\, w_1^{-1}}$ and then set

$$\left.\begin{array}{l} F = \\ \text{or} \\ \overline{F} = \end{array}\right\} \widehat{w\, w_1^{-1}} \cdot F_1 \cdot z_1 z^{-1}$$

whichever makes F analytic. In the first case

$$\begin{aligned} \varphi F z &= \varphi \widehat{w w_1^{-1}} \cdot F_1 \circ z_1 \text{ which, by (1.4.1),} \\ &= \varphi w w_1^{-1} \varphi F_1 z_1 \\ &= \varphi w w_1^{-1} w_1 f \\ &= \varphi w f \\ &= w f, \end{aligned}$$

and the same proof works in the second case. Consider the following diagram:

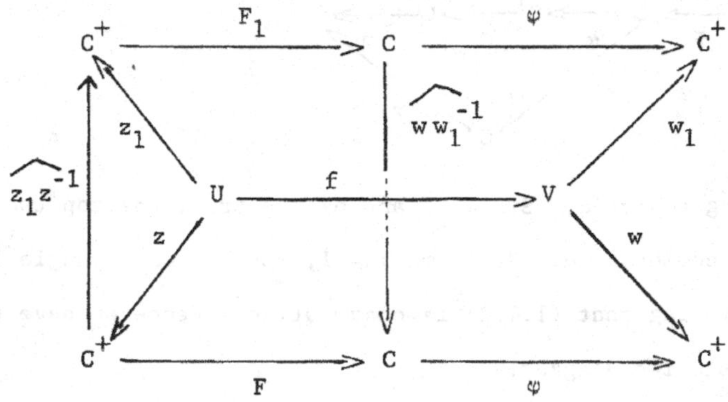

Since $\varphi \cdot w w_1^{-1} = w \cdot w_1^{-1} \circ \varphi$, and F analytic, we see that F is unique, proving the lemma.

Theorem 1.4.3. Let $\mathfrak{X}$, $\mathfrak{Y}$, and $\mathfrak{T}$ be Klein surfaces, let f and g be a continuous map of X into Y and Y into T respectively such that $f(\partial X) \subset \partial Y$ and $g(\partial Y) \subset \partial T$. If any two of the following statements is true, so is the third:

a) $f : \mathfrak{X} \longrightarrow \mathfrak{Y}$ is a non-constant morphism,

b) $g : \mathfrak{Y} \longrightarrow \mathfrak{T}$ is a non-constant morphism,

c) $g f : \mathfrak{X} \longrightarrow \mathfrak{T}$ is a non-constant morphism.

Proof. Let $x \in X$, and let (U,z), (V,w), and (S,ζ) be dianalytic charts at x, $f(x)$, and $gf(x)$ respectively, such that $f(U) \subset V$, $g(V) \subset S$.

Assume first that a) and b) are true; then by lemma 1.4.2, we have the following diagram,

(1.4.2)

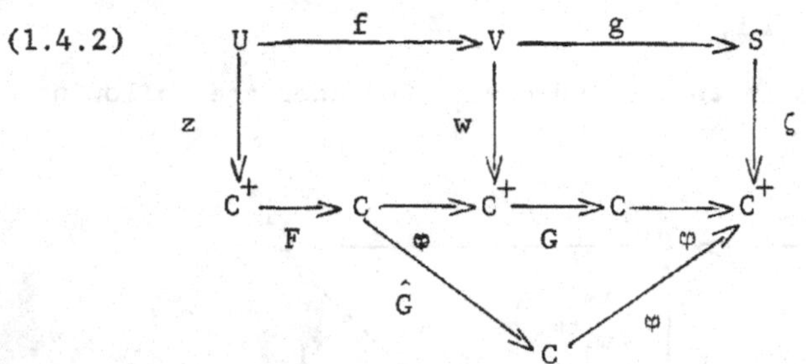

F and G being analytic. Since a) and b) are true, the top two rectangles are commutative. By lemma 1.4.1, the bottom triangle is commutative; proving that (1.4.2) is commutative. Hence we have the following commutative diagram:

(1.4.3)

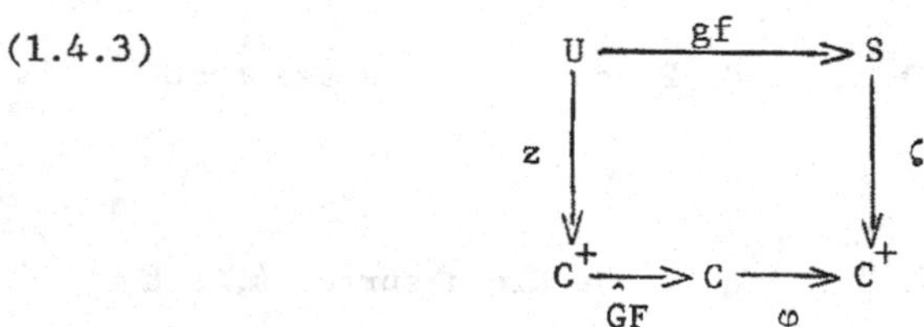

proving that c) is true.

Assume that b) and c) are true; then we have the following diagram,

(1.4.4)

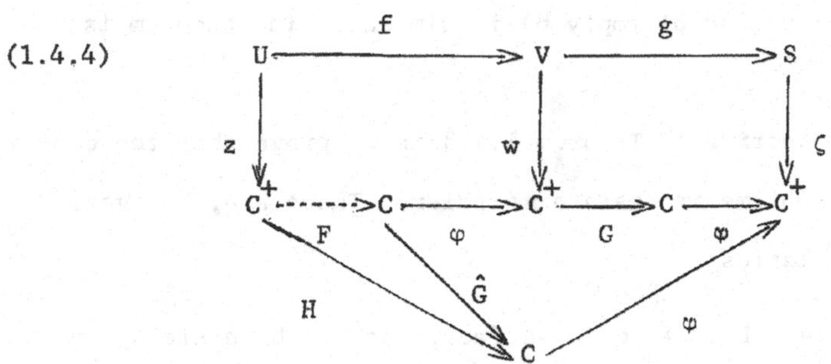

H and G being analytic. Our goal is to find an analytic function F making the diagram commutative. We see that a natural candidate for F is $\hat{G}^{-1}H$. Unfortunately $\hat{G}^{-1}$ need not exist. Since wfz^{-1} is a continuous map on $z(U)$, if F exists, φF is forced. If in addition, F is analytic, it is unique. Since $\hat{G}$ is a non-constant analytic function it will have local inverses, except on a discrete subset D of $A \equiv w(V) \cup \overline{w(V)}$. Assume first that $wf(x) \notin D$. Then U, V, and S may be shrunk to U', V', and S' consistent with the conditions above, such that $w(V') \cup \overline{w(V')} \equiv A'$ does not meet D. On A', $\hat{G}$ has well defined local inverses $(\hat{G}_j^{-1})_{j\in J}$ and each $\hat{G}_j^{-1}$ is analytic. Let $F_j \equiv \hat{G}_j^{-1}H$. By lemma 1.4.1, $\varphi G \varphi = \varphi \hat{G}$; thus $\varphi G \varphi \hat{G}_j^{-1} = \varphi$. Let F_j play the role assigned to F in diagram (1.4.4), and note that then it is commutative. Since F_j is analytic it is unique, as remarked above. Thus F is a well defined dianalytic function on $z(U)$ minus a discrete subset D'. Since φF must be wfz^{-1} on $z(U) - D'$, and since wfz^{-1} is continuous on $z(U)$, the singularities of F on D' are removable. Let them be removed. Thus F exists making (1.4.4) commutative. Hence a) is true.

The proof that c) and a) imply b) is similar. The theorem is proved.

The primary objective of Theorem 1.4.3 is to prove that the composition of two morphisms is again a morphism. There are, however, several other corollaries.

Corollary 1.4.4. Let X be a surface, let $\mathfrak{Y}$ be a Klein surface, and let $f : X \longrightarrow Y$ be a continuous map. Then there exists at most one dianalytic structure on X making f into a morphism.

Proof. Let $\mathfrak{X}_1$ and $\mathfrak{X}_2$ be dianalytic structures on X such that $f : \mathfrak{X}_i \longrightarrow \mathfrak{Y}$ is a morphism, $i = 1, 2$. Using the identity map 1 on X to map $\mathfrak{X}_1$ to $\mathfrak{X}_2$, and applying Theorem 1.4.3 to $X_1 \xrightarrow{1} X_2 \xrightarrow{f} Y$ we see that $1 : \mathfrak{X}_1 \longrightarrow \mathfrak{X}_2$ is a morphism. Since the argument is symmetric, $\mathfrak{X}_1 = \mathfrak{X}_2$, proving the corollary.

Corollary 1.4.5. Let $\mathfrak{X}$ be a Klein surface, let Y be a surface, and let $f : X \longrightarrow Y$ be a continuous map. Then there exists at most one dianalytic structure on Y making f into a morphism.

The proof of (1.4.5) is a close analogue to that of (1.4.4).

Let $E_0(\mathfrak{X})$ denote the relative algebraic closure of R in $E(\mathfrak{X})$, Then either $E_0(\mathfrak{X}) = R$, or it is R-isomorphic to C. Given $\underline{f} \in E(\mathfrak{X})$, we will denote the corresponding map from X to $\mathrm{cl}\, C^+$ by f. If $\underline{f} \notin E_0(\mathfrak{X})$, then f is clearly a non-constant map.

The set of non-constant morphisms from $\mathfrak{X}$ to $\mathfrak{Y}$ will be denoted by $\mathrm{Mor}\,(\mathfrak{X},\mathfrak{Y})$.

Theorem 1.4.6. Given $\underline{f} \in E(\mathfrak{X})$, then $f : \mathfrak{X} \longrightarrow \mathrm{cl}\, C^+$ is a morphism of Klein surfaces, f being the map of $\underline{f}$. Further,

$\underline{f} \longmapsto f$ induces a bijection from $E(\mathfrak{X}) - E_0(\mathfrak{X})$ to Mor $(\mathfrak{X}, \mathrm{cl}\ C^+)$.

Proof. Let $\underline{f} \in E(\mathfrak{X}) - E_0(\mathfrak{X})$, let $x \in U \subset X$ with (U,z) a dianalytic chart, and let f_U be the corresponding function of $\underline{f}$. If $\underline{f}(x) \neq \infty$, we may choose U so that f has no pole on U, while if $f(x) = \infty$, choose U so that $\frac{1}{\underline{f}}$ has no pole on U. Correspondingly choose V in $\mathrm{cl}\ \mathbb{C}^+$ to be either $\mathbb{C}^+$ or $\mathrm{cl}\ \mathbb{C}^+ - \{0\}$, and let w be either the identity map or given by $w(c) = \overline{1/c}$. Then we have the commutative diagram

(1.4.5)
$$\begin{array}{ccccc} U & & \xrightarrow{\quad f \quad} & & V \\ \downarrow{\scriptstyle z} & & & & \downarrow{\scriptstyle w} \\ \mathbb{C}^+ & \xrightarrow[F]{} & \mathbb{C} & \xrightarrow[\varphi]{} & \mathbb{C}^+ \end{array}$$

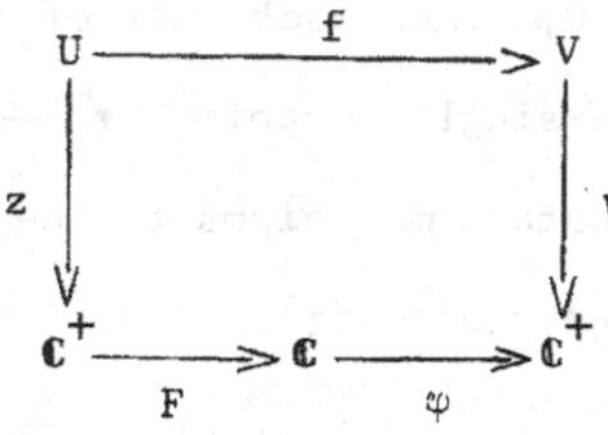

where $F = f_U z^{-1}$ or $F = 1/f_U z^{-1}$. Hence f is a morphism.

Now let $\underline{f}, \underline{f}' \in E(\mathfrak{X}) - E_0(\mathfrak{X})$ and assume $f = f'$. Then for every dianalytic chart (U,z), with U connected, we have $f_U = f'_U$ or $f_U = \overline{f'_U}$. Since both $f_U z^{-1}$ and $f'_U z^{-1}$ are non-constant and analytic, $f_U = f'_U$ and thus $\underline{f} = \underline{f}'$.

Let g be a non-constant morphism of $\mathfrak{X}$ into $\mathrm{cl}\ C^+$. For $x \in X$, choose a dianalytic chart (U,z) with either $g(U) \subset \mathbb{C}^+ = V_1$ or $g(U) \subset \mathrm{cl}\ \mathbb{C}^+ - \{0\} = V_2$, and let w_1 be the inclusion map, $w_2(p) = \overline{1/p}$. Then we have a commutative diagram

(1.4.6)
$$\begin{array}{ccccc} U & & \xrightarrow{\quad g \quad} & & V_j \\ \downarrow{\scriptstyle z} & & & & \downarrow{\scriptstyle w_j} \\ \mathbb{C}^+ & \xrightarrow[G]{} & \mathbb{C} & \xrightarrow[\varphi]{} & \mathbb{C}^+ \end{array}$$

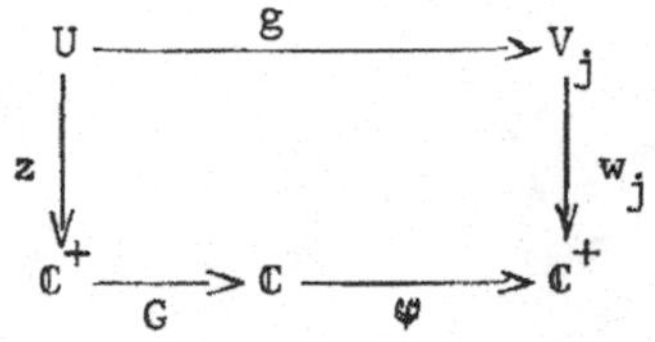

either for $j = 1$ or $j = 2$, where G is analytic. Define g_U by

$$g_U = \begin{cases} G \cdot z & j = 1 \\ \dfrac{-1}{G \cdot z} & j = 2. \end{cases}$$

To conclude the proof, we must show that g_U does not depend on the choice of j, and that (g_U) defines a function $\underline{g} \in E(\mathfrak{X})$, which induces the morphism g. This is easily done.

Remark. If $E_0(\mathfrak{X}) \cong \mathbb{C}$, then each pair of conjugate constant functions corresponds to a single morphism $\mathfrak{X} \longrightarrow \mathrm{cl}\ \mathbb{C}^+$. If $\partial X = \emptyset$ but $E_0(\mathfrak{X}) = \mathbb{R}$, then constant morphisms to the interior of $\mathrm{cl}\ \mathbb{C}^+$ do not correspond to elements of $E(\mathfrak{X})$.

Let $f : \mathfrak{X} \longrightarrow \mathfrak{Y}$ be a morphism of Klein surfaces. We now define the "backwards map" of meromorphic function fields.

Theorem 1.4.7. Given $f : \mathfrak{X} \longrightarrow \mathfrak{Y}$, non-constant, there is a unique $\mathbb{R}$ - monomorphism $f^* : E(\mathfrak{Y}) \longrightarrow E(\mathfrak{X})$ with the property that the map of $f^*\underline{g}$ is $g f$ (where g is the map of $\underline{g}$), for all $\underline{g} \in E(\mathfrak{Y})$.

Proof. Let $\underline{g} \in E(\mathfrak{Y})$, let $x \in X$, and let (U,z), (V,w) be dianalytic charts such that $f(U) \subset V$, $x \in U$, and let F be that analytic function on $z(U)$ which makes the following diagram commute.

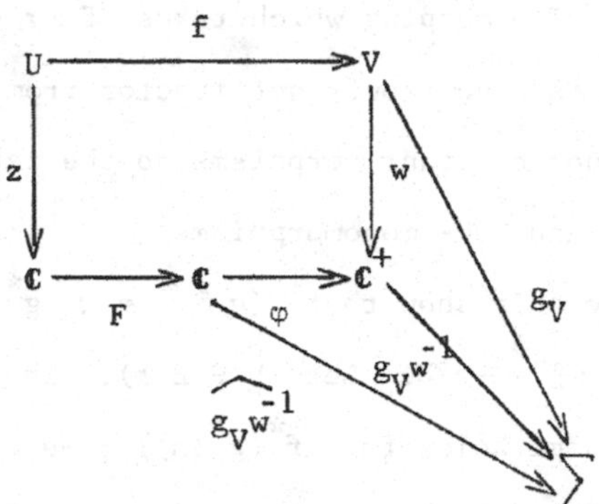

Set $h_U = \widehat{g_V w^{-1}} \cdot F \cdot z$. It is easily checked that (h_U) defines a meromorphic function $\underline{h} \in E(\mathfrak{X})$. Further, using (1.4.1),

$$\begin{aligned} \varphi\, h_U &= \varphi\, g_V w^{-1}\, \varphi\, F\, z \\ &= \varphi\, g_V w^{-1}\, w\, f \\ &= \varphi\, g_V f \\ &= g\, f. \end{aligned}$$

It is now immediate that, by setting $f^* \underline{g} = \underline{h}$, we obtain an $\mathbb{R}$-monomorphism from $E(\mathfrak{Y})$ to $E(\mathfrak{X})$ with the desired properties.

If $\underline{g} \in E(\mathfrak{Y}) - E_0(\mathfrak{Y})$, then $f^*\underline{g}$ is uniquely determined by the fact that its map is $g\,f$. If $\underline{g}$ is constant, represent it as the sum of two non-constant functions. (This can actually be done globally -- see 1.6.6 -- but for the purpose of this proof it suffices to do it locally.)

The following corollary also has a direct topological proof.

Corollary 1.4.8. Let $f : \mathfrak{X} \longrightarrow \mathfrak{Y}$ be a morphism of Klein surfaces. If Y is orientable and without boundary then the same is true of X.

Proof. $E_0(\mathfrak{Y}) \cong C$ and $f^*(E_0(\mathfrak{Y})) \subseteq E(\mathfrak{X})$.

Theorem 1.4.9. The mapping which takes $f : \mathfrak{X} \longrightarrow \mathfrak{Y}$ to $f^* : E(\mathfrak{Y}) \longrightarrow E(\mathfrak{X})$ is a controvariant functor from the category of Klein surfaces and non-constant morphisms to the category of extension fields of $\mathbb{R}$ and $\mathbb{R}$ - monomorphisms.

Proof. We need only show that $(g\,f)^* = f^*g^*$, where $f : \mathfrak{X} \longrightarrow \mathfrak{Y}$ and $g : \mathfrak{Y} \longrightarrow \mathfrak{T}$. Let $\underline{h} \in E(\mathfrak{T})$. The map of $(gf)^*\underline{h}$ is $h\,g\,f$, and the same holds for $f^*(g^*(h))$. We now use the uniqueness part of (1.4.7) to prove the theorem.

Remark. We will see in Chapter 2 that if we restrict ourselves to compact Klein surfaces, and function fields in one variable over R, then the functor above has an inverse.

§5 Existence of Dianalytic Structure

Let $f : \mathfrak{X} \longrightarrow \mathfrak{Y}$ be a morphism of Klein surfaces. Let $x \in X$ and let (U,z) and (V,w) be dianalytic charts at x and $f(x)$ respectively such that $z(x) = 0 = w(f(x))$, $f(U) \subset V$, and such that $f|U$ is $w^{-1}\varphi F z$ if $f(x) \in \partial Y$ and $f|U = w^{-1}F z$, where F is analytic. Since $\hat{F}(0) = 0$ and since $\hat{F}$ is analytic it has a convergent power series representation $\sum_{j=e}^{\infty} a_j z^j$, where $e \geq 1$ and $a_e \neq 0$. The integer e is called the ramification index of f at x. We say that f is ramified at x if $e > 1$; otherwise we say f is unramified at x. Clearly e is independent of the choice of (U,z) and (V,w). One immediately notes that classical planar function theory implies that the set of all ramified points of f in X is discrete.

The main theorem of this section is an existence theorem for dianalytic structures, which applies to a large class of surfaces, including in particular all compact surfaces. But first a lemma which is a consequence of the Schwarz Reflection Principle.

Lemma 1.5.1. Let A be an open set in C^+ and let $f : A \longrightarrow C^+$ be a dianalytic homeomorphism. Let B be an open subset of $\varphi^{-1}(A)$ (= A union the image of A under conjugation), and let $g : B \longrightarrow C$ be a homeomorphism such that

$$\varphi g = f\varphi \text{ on } B. \tag{1.5.1}$$

Then g is dianalytic.

Proof. Since f is a homeomorphism of $A \subset C^+$ into C^+, $f(z) \in R$ if and only if $z \in R$; thus formula (1.5.1) implies that

$g(z) \in R$ if and only if $z \in R$. Now let z and $\bar{z} \in B$, with say $z \in C^+$. Then $\varphi g(z) = f\varphi(z) = f(z) = f\varphi(\bar{z}) = \varphi g(\bar{z})$, so $g(\bar{z}) = \overline{g(z)}$. Let G be the interior of $B \cap C^+$ in C. Since $\varphi g \mid G = f \mid G$, and since f and g do not assume real values on G, either $g \mid G = f \mid G$ or $g \mid G = \bar{f} \mid G$; thus g is dianalytic on G. Using the classical reflection principle of Schwarz, we see that g is dianalytic on B, proving the lemma.

Turning our attention to the main theorem of the section, let X be a two-manifold, let $\mathfrak{Y}$ be a Klein surface, and let $f : X \longrightarrow Y$ be a continuous map such that $f(\partial X) \subset \partial Y$. We want to put a dianalytic structure on X in terms of which f will be a morphism of Klein surfaces. The next theorem asserts that if f is a morphism locally, then it is a morphism globally.

Theorem 1.5.2. Let X be a surface, $\mathfrak{Y}$ a Klein surface, and let f be a non-constant continuous map of X into Y such that $f(\partial X) \subset \partial Y$. Assume that there exist the following: an atlas $\underset{\sim}{U} \equiv (U_j, z_j)_{j \in J}$ of X, a dianalytic atlas $\underset{\sim}{V} \equiv (V_j, w_j)_{j \in J}$ of $\mathfrak{Y}$ for which $f(U_j) \subset V_j$, and analytic functions F_j on $z_j(U_j)$ such that

$$(1.5.2) \qquad f \mid U_j = w_j^{-1} \varphi F_j z_j, \quad \text{for all } j \in J.$$

Then $\underset{\sim}{U}$ is a dianalytic atlas on X. If $\mathfrak{X}$ is the dianalytic structure containing $\underset{\sim}{U}$, then f is a morphism from $\mathfrak{X}$ to $\mathfrak{Y}$ and $\mathfrak{X}$ is the only dianalytic structure on X making f a morphism.

Proof. If necessary, replace $\underset{\sim}{U}$ with a refinement, so that, given $j, k \in J$, $j \neq k$, $f \mid U_j \cap U_k$ is a local homeomorphism. We must show that $z_j z_k^{-1}$ is dianalytic on $z_k(U_j \cap U_k)$. We can cover $U_j \cap U_k$ by open sets $(T_m)_{m\in M}$ such that F_j and F_k are homeomorphisms when restricted to $z_j(T_m)$ and $z_k(T_m)$, respectively. Note that $z_j z_k^{-1}$ is real-valued on $z_k(\partial X \cap U_j \cap U_k)$ and hence, in view of the Schwarz Reflection Principle, it suffices to check dianalyticity on $z_k(U_j \cap U_k - \partial X)$: i.e., it suffices to show that $z_j z_k^{-1}$ is dianalytic on each $z_k(T_m - \partial X)$. Let $T_m - \partial X = T_m'$. Note that $w_j^{-1} \varphi F_j z_j \mid T_m' = f \mid T_m' = w_k^{-1} \varphi F_k z_k \mid T_m'$. Since w_j, F_k, and z_k are homeomorphisms when so restricted, we see that $\varphi F_j z_j z_k^{-1} F_k^{-1} = w_j w_k^{-1} \varphi$ on $F_k z_k(T_m')$. Since $w_j w_k^{-1}$ is dianalytic on $w_k(T_m')$, we can invoke (1.5.1) and conclude that $F_j z_j z_k^{-1} F_k^{-1}$ is dianalytic when so restricted, which immediately implies that $z_j z_k^{-1} \mid T_m'$ is dianalytic; thus $\underset{\sim}{U}$ is a dianalytic coordinate covering of X. To see that f is a morphism, one merely needs to check the definition. Uniqueness is an immediate consequence of (1.4.4).

Lemma 1.5.3. Let F be an analytic function on some neighborhood A of 0 and assume that $\sum_{j=n}^{\infty} a_j z^j$ is the power series expansion of F at 0, each a_j being real and $a_n > 0$, where $n \geq 1$. There exists an injective analytic function G on some subneighborhood B of A whose power series has real coefficients, such that $F = G^n$.

Proof. Let $F(z)/z^n \equiv H(z)$ for all $z \in A - \{0\}$. Then H has a removable singularity at 0; let it be removed. Let $A' \subset A$ be a simply connected neighborhood of 0, such that $A' \cap R$ is connected and H has no zeroes on A'. There exists an analytic function h on A' such that $H(z) = e^{h(z)}$ for all $z \in A'$. Since all $a_j \in R$, F is real on $A \cap R$; thus so is H. Let $h(z) = u(z) + i\,v(z)$ for all $z \in A'$, u and v being real-valued functions. Clearly $H(0) = a_n$. Since $a_n > 0$, $v(0)$ is congruent to zero mod 2π. By adding a complex constant to $h(z)$, we may assume that $v(0) = 0$; thus $v(z) = 0$ for all $z \in A' \cap R$. Let $G(z) \equiv z\,e^{h(z)/n}$ for all $z \in A'$. Note that on $A' \cap R$, G is real-valued and that $G^n = F$. Since $G - G(0)$ has a simple zero at 0, it is injective on some sub-neighborhood B of A'.

Let $f : \mathfrak{X} \longrightarrow \mathfrak{Y}$ be a morphism of Klein surfaces, let $x \in \partial X$, and let e be the ramification index of f at x. Let us choose dianalytic charts (U,z) of x and (V,w) of $f(x)$ such that $z(x) = 0 = w(f(x))$, for which there exists an analytic function F such that $f\,|\,U = w^{-1}\varphi F z$. Note that $F(z(U) \cap R) \subset R$; thus the power series development of F at 0, $\sum_{j=e}^{\infty} a_j z^j$, has real coefficients. Then we have either $F = G^e$ or $F = -G^e$, depending on whether $a_e > 0$ or $a_e < 0$.

Let U_1 be a subdomain of U such that $z(U_1) = B$, of the lemma. If $z_1 = G\,z$, then $f\,|\,U_1$ is $w^{-1}\varphi\, z_1^{\,e}$ or $w^{-1}\varphi\, {-z_1^{\,e}}$. In this event, we will say that (U_1, z_1) is in <u>normal form</u> for f and (V,w) at x.

Let $f : \mathfrak{X} \longrightarrow \mathfrak{Y}$ be a morphism of Klein surfaces. Let I be a closed interval in ∂Y, let I_1 and I_2 be two intervals in ∂X, both mapped homeomorphically by f onto I, that are either disjoint, intersect at one endpoint, or intersect at two endpoints. Let X' be the space obtained from X by identifying I_1 and I_2 via f, and let $q : X \longrightarrow X'$ be the quotient map. Then we have the following commutative diagram of spaces and continuous surjections:

(1.5.4)

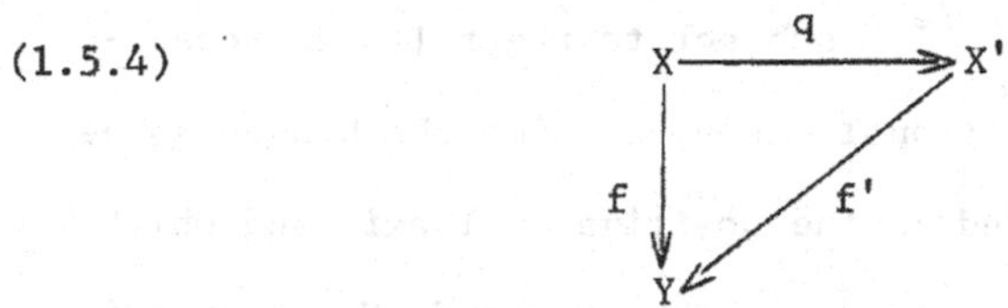

Theorem 1.5.4. There exists a unique dianalytic structure $\mathfrak{X}'$ on X' under which f' is a morphism.

Proof. Since we will appeal to (1.5.2), we must define an atlas $\underline{U}' \equiv (U_j', z_j')_{j \in J}$ on X' with the requisite properties. In view of the fact that X' is obtained from X by means of identification, we need only find such charts at points $x' \in I' \equiv q(I_1) = q(I_2)$. In order to do this we need to consider three cases, where $y \equiv f'(x')$:

1. y is an interior piont of I;
2. y is an end point of I, and $x' \in \partial X'$; and
3. y is an end point of I, and $x' \notin \partial X'$.

Consider first Case 1. $q^{-1}(x') = \{x_1, x_2\}$, where $x_j \in I_j$, $j = 1, 2$. Since x_j is an interior point of I_j, the ramification index e_j of f at x_j must be odd. Let (V,w) be a chart at y such that $w(y) = 0$. In view of the remarks about normal forms, it is clear that we can choose dianalytic charts (U_k, z_k) of x_k, $k = 1, 2$, so that $f \mid U_1 = w^{-1} \varphi z_1^{e_1}$ and $f \mid U_2 = w^{-1} \varphi (-z_2)^{e_2}$. (Note: Since e_1 is odd G_1 may be found so that $F_1 = G_1^{e_1}$. Similarly G may be found so that $F_2 = G^{e_2}$; let $G_2 = -G$.) Let $e = e_1 + e_2$ and note that since e_1 and e_2 are odd, e is even. Let U' be a neighborhood of x' in $q(U_1 \cup U_2)$, and define $z' : U' \longrightarrow C$ as follows: on $q(U_1)$, let z' be $(z_1 q^{-1})^{2e_1/e}$, on $q(U_2)$, let z' be $\exp(2\pi i\, e_1/e)(z_2 q^{-1})^{2e_2/e}$, subject to the following conditions. When dealing with fractional exponents, let the branch above be the one which is real-valued on the positive real axis and which is continuous in C^+. Since z_1 and $-z_2$ map U_1 and U_2 respectively into C^+, $z' \mid q(U_1)$ and $z' \mid q(U_2)$ are well defined. To show that z' is well defined let $t' \in U' \cap I'$ and let $q^{-1}(t') = \{t_1, t_2\}$, where $t_j \in I_j$, $j = 1, 2$. Since $f = f'q$, $f(t_1) = f(t_2)$; thus $\varphi(z_1(t_1))^{e_1} = w\, f(t_2) = \varphi(-z_2(t_2))^{e_2}$. Since $t' \in I'$, $z_1(t_1)$ and $z_2(t_2)$ are in R. As a consequence $(z_1(t_1))^{e_1} = (-z_2(t_2))^{e_2} = -(z_2(t_2))^{e_2}$. Since both e_1 and e_2 are odd, $z_1(t_1) \geq 0$ if and only if $z_2(t_2) \leq 0$. Let $z_j(t_j) = r_j \exp i\,\theta_j$, $j = 0, 1$, θ_j being either 0 or π. In case $\theta_1 = 0$, $\theta_2 = \pi$; then $(z_1(t_1))^{2e_1/e} = r_1^{2e_1/e}$, and

$$(\exp 2\pi i\, e_1/e)(z_2(t_2))^{2e_2/e} = (\exp 2\pi i\, e_1/e)(\exp 2\pi i\, e_2/e)\, r_2^{2e_1/e}$$
$$= r_2^{2e_1/e}.$$

Since $r_1^{e_1} = r_2^{e_2}$, as noted above, $(z_1(t_1))^{2e_1/e} = (\exp 2\pi i\, e_1/e)\cdot(z_2(t_2))^{2e_2/e}$. In case $\theta_1 = \pi$, then $\theta_2 = 0$. $(z_1(t_1))^{2e_1/e} = (\exp 2\pi\, e_1/e)\, r_1^{2e_1/e}$, and $\exp(2\pi i\, e_1/e)(z_2(t_2))^{2e_2/e} = \exp(2\pi i\, e_1/e)\, r_2^{2e_2/e}$, proving that z' is well defined in this case. Since z' is continuous on $q(U_1)$ and on $q(U_2)$, it is continuous. Now we wish to show that z' is injective. Let j be

Case 1

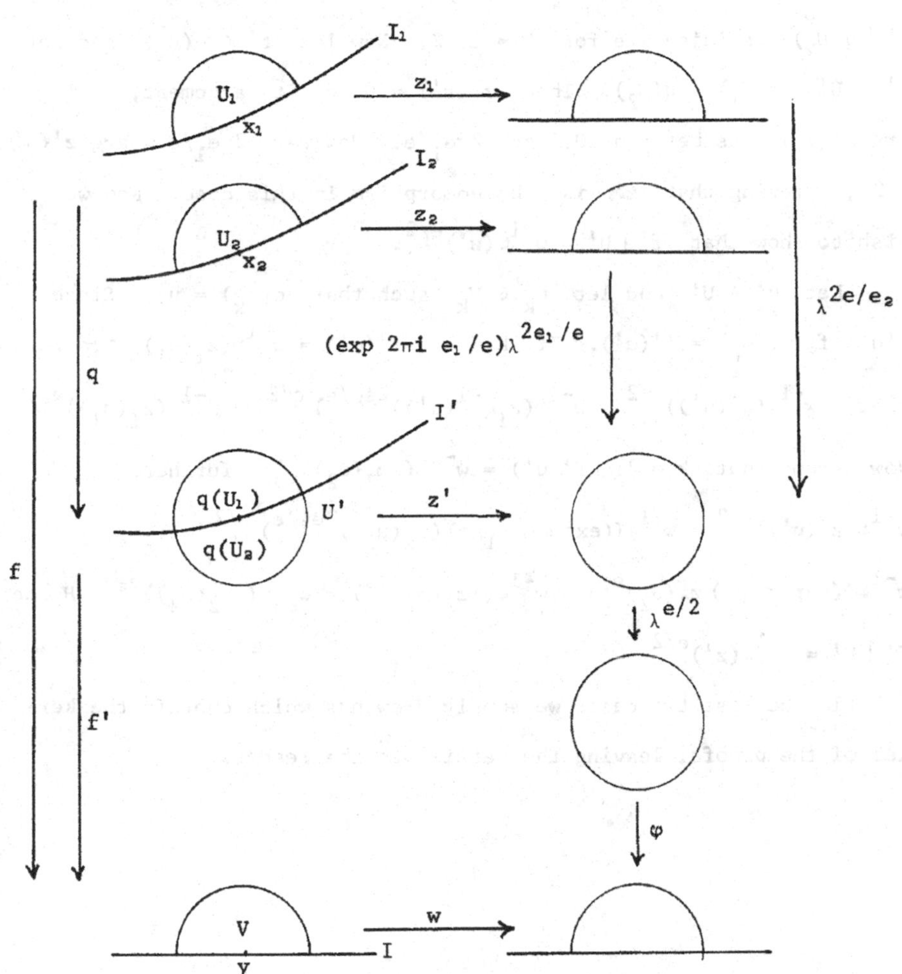

I₁
U₁
x₁
z₁
I₂
U₂
x₂
z₂
λ^{2e/e₂}
(exp 2πi e₁/e)λ^{2e₁/e}
q
I'
q(U₁)
q(U₂)
U'
z'
f
λ^{e/2}
f'
φ
V
y
w
I

such that $e_j \leq e_{1-j}$. Then, $2e_j \leq e_1 + e_2 \leq 2e_{1-j}$ and $0 \leq 2e_j/e \leq 1 \leq 2e_{1-j}/e$. Further, $2e_1/e + 2e_2/e = 2$; thus $2e_{1-j}/e < 2$. Clearly the map $z \longmapsto (z)^{2e_k/e}$ maps C^+ injectively into C; thus $z' \mid q(U_k)$ is injective for $k = 1, 2$. Now let $t' \in q(U_1)$ and let $u' \in U' - q(U_1) \subset q(U_2)$. Then $z'(t') = 0$ or its argument, $\arg z'(t')$, is between 0 and $2\pi e_1/e$. However $2\pi e_1/e < \arg z'(u') < 2\pi$, proving that z' is a homeomorphism in this case. Now we wish to show that $f' \mid U' = w^{-1}\varphi(z')^{e/2}$.

Let $u' \in U'$ and let $u_k \in U_k$ such that $q(u_k) = u'$. Since $f'q = f$, $f(u_k) = f'(u')$. If $k = 1$, $f'(u') = w^{-1}\varphi(z_1(u_1))^{e_1}$; further, $w^{-1}\varphi(z'(u'))^{e/2} = w^{-1}\varphi(z_1 q^{-1}(u'))^{2e_1/e)\, e/2} = w^{-1}\varphi(z_1(u_1))^{e_1}$. Now assume that $k = 2$. $f'(u') = w^{-1}\varphi(-z_2(u_2))^{e_2}$; further, $w^{-1}\varphi(z'(u'))^{e/2} = w^{-1}\varphi((\exp 2\pi i\, e_1/e)\,(z_2(u_2))^{2e_2/e)\, e/2} = w^{-1}\varphi((\exp \pi i\, e_1)\, z_2(u_2)^{e_2}) = w^{-1}\varphi(-z_2(u_2)^{e_2}) = w^{-1}\varphi(-z_2(u_2))^{e_2}$. Hence $f' \mid U' = w^{-1}\varphi(z')^{e/2}$.

In the last two cases we supply drawings which contain the kernel of the proofs, leaving the details to the reader.

Case 2

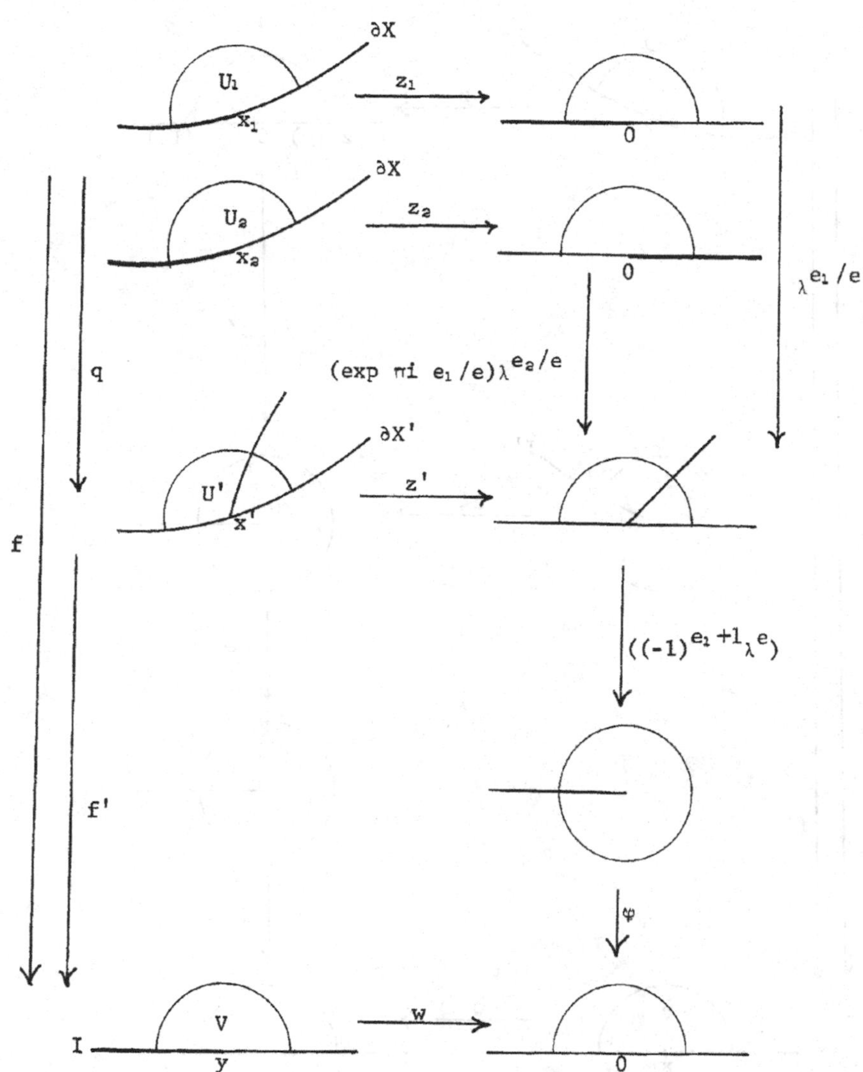
∂X
U_1
x_1
z_1
0
∂X
U_2
x_2
z_2
0
$\lambda^{e_1/e}$
q
$(\exp \pi i\ e_1/e)\lambda^{e_2/e}$
∂X'
U'
x'
z'
f
$((-1)^{e_1+1}\lambda^e)$
f'
φ
V
w
I
y
0

Case 3

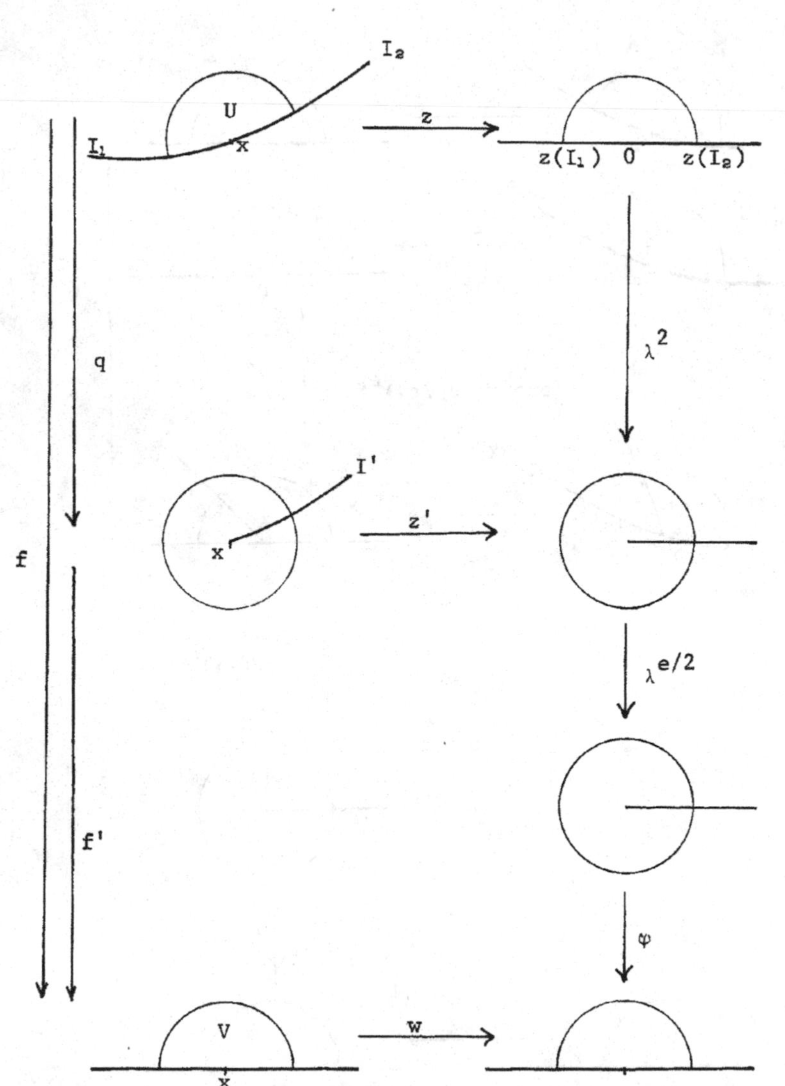

§6 Unramified Double Covers

A morphism $f : \mathfrak{X} \longrightarrow \mathfrak{Y}$ of Klein surfaces will be called double cover if each $y \in Y$ has a neighborhood V such that either $f^{-1}(V)$ has two components, each of which is mapped homeomorphically onto V by f; or $f^{-1}(y) = \{x\}$, and there exist dianalytic charts (U,z), and (V,w) of x and y respectively such that $z(x) = 0 = w(y)$, $f(U) \subset V$, and

$$wf\,|\,U = \begin{cases} \varphi z & \text{if } y \in \partial Y \text{ and } x \notin \partial X, & (1.6.a)\\ \varphi z^2 & \text{if } y \in \partial Y \text{ and } x \in \partial X, & (1.6.b)\\ z^2 & \text{if } y \notin \partial Y; & (1.6.c)\end{cases}$$

φ being the folding map; f is unramified if (1.6.b) and (1.6.c) never occur.

The goal of this section is to show the existence and uniqueness of three double covers. These covers will be used subsequently in order to apply results about Riemann surfaces to Klein surfaces.

Theorem 1.6.1. There exists a double cover $f : \mathfrak{X}_C \longrightarrow \mathfrak{X}$ of $\mathfrak{X}$ by a Riemann surface $\mathfrak{X}_C$ (here we allow X_C to be disconnected), such that $\mathfrak{X}_C$ has an antianalytic involution σ with $f\sigma = f$. If $(\mathfrak{X}_C', f', \sigma')$ is any other such triple, then there is a unique analytic isomorphism $\rho : \mathfrak{X}_C' \longrightarrow \mathfrak{X}_C$ such that $f' = f\rho$.

Further, f is unramified, and σ is the only antianalytic automorphism of $\mathfrak{X}_C$ such that $f\sigma = f$.

Proof. $\mathfrak{X}_C$ could be constructed by using the methods of §5, but it is easier to procede directly. Let $(U_j, z_j)_{j\in J}$ be a dianalytic atlas of $\mathfrak{X}$. For each $j \in J$, let $U_j' \equiv U_j \equiv U_j'', z_j' = z_j$, and $z_j'' \equiv \bar{z}_j$. Let Ω be the disjoint union of the U_j' 's and the U_j'' 's. Let us now proceed to make identification of two types.

(1) If W is a component of $U_j \cap U_k$ and if $z_j z_k^{-1}$ is analytic (resp. antianalytic) on $z_k(W)$, then identify its image in U_j' with its image in U_k' (resp. its image in U_j' with its image in U_k'') and its image in U_j'' with its image in U_k'' (resp. its image in U_j'' with its image in U_k').

(2) Let $B_j \equiv \partial X \cap U_j$ and identify its image in U_j' with its image in U_j''. ¶Let X_C be the quotient space of Ω, with all the above identifications, and let q be the quotient map. Let $\tilde{U}_j$ be the image of $U_j' \cup U_j''$ in X_C, and let $\tilde{z}_j$ map $\tilde{U}_j$ into C as follows: $\tilde{z}_j \mid U_j' = z_j'$ and $\tilde{z}_j \mid U_j'' = z_j''$. It is easily seen that $\tilde{z}_j$ is a homeomorphism. Using the Schwarz reflection principal, we can see that $\tilde{z}_k \tilde{z}_j^{-1}$ is analytic on $\tilde{z}_j(\tilde{U}_j)$; thus $(\tilde{U}_j, \tilde{z}_j)_{j \in J}$ is an analytic atlas of X_C. Let $\mathfrak{X}_C$ be the associated analytic structure on X_C. Let $f : X_C \longrightarrow X$ be induced by the identity maps $U_j' \longrightarrow U_j$ and $U_j'' \longrightarrow U_j'$. Clearly $f\sigma = f$.

Let σ' be any antianalytic automorphism of $\mathfrak{X}_C$ with $f\sigma' = f$. Suppose there is an $x \in X_C$ with $y \equiv \sigma\sigma'(x) \neq x$. Then y is on the deck above x and hence $\sigma\sigma'(x) = \sigma(x)$. By continuity $\sigma\sigma'$ is analytic, both being non-constant; which is absurd. Hence $\sigma' = \sigma$, and we see that σ is the only antianalytic automorphism of $\mathfrak{X}_C$ with $f \doteq f\sigma$.

We now show that $(\mathfrak{X}_C, f, \sigma)$ is unique. Let $(\mathfrak{X}_C', f', \sigma')$ be another such triple. Suppose there is an analytic map $\rho : \mathfrak{X}_C' \longrightarrow \mathfrak{X}_C$

such that $f\rho = f'$. Suppose ρ is not surjective. Then $\mathfrak{X}_C$ must have two components and ρ must be a surjection onto one of them. Hence $\rho = \rho\sigma'$, which is antianalytic, which is impossible. Since both f and f' are double coverings, ρ is an isomorphism.

Hence the proof of 1.6.1 will be completed by proving the following proposition.

Proposition 1.6.2. Let $\mathfrak{X}$ be a Klein surface, $\mathfrak{Y}$ a Riemann surface (with $\partial Y = \emptyset$), and let $g : \mathfrak{Y} \longrightarrow \mathfrak{X}$ be a non-constant morphism. Then there is a unique analytic map $\rho : \mathfrak{Y} \longrightarrow \mathfrak{X}_C$ such that $f\rho = g$.

Proof. Let (V,w) be an analytic chart on $\mathfrak{Y}$, and (U,z) a dianalytic chart on $\mathfrak{X}$ with $g(V) \subset U$, such that there exists an analytic function G on $w(V)$ with $g|V = z^{-1}\varphi G w$. We have $\tilde{U}$ and $\tilde{z}$ as in the construction of $\mathfrak{X}_C$, with $\tilde{U} = f^{-1}(U)$ and $\tilde{z} : \tilde{U} \longrightarrow C$. Hence the following diagram commutes:

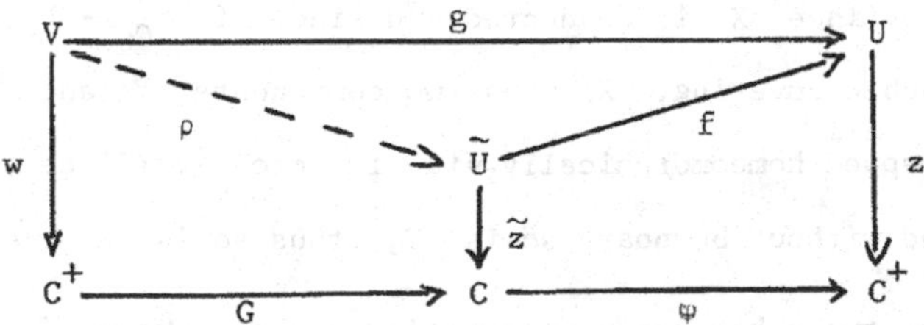

and we define ρ on V by $\rho = \tilde{z}^{-1} G w$ so that, a simple diagram chase yields, $f\rho = g$. If ρ is well defined on $\mathfrak{Y}$, then it is an analytic map. Let $(V_1 w_1)$, (U_1, z_1) and G_1 also satisfy the above, and set $\rho_1 = \tilde{z}_1^{-1} G_1 w_1$. Assume that there is a $y \in V \cap V_1$ with

$\rho(y) \neq \rho_1(y)$. Then $\rho(y) = \sigma\rho_1(y)$, and by continuity there is a neighborhood N of y such that $\rho \mid N = (\sigma\rho_1) \mid N$. But ρ is analytic and $\sigma\rho_1$ is antianalytic on N, so this is impossible. Hence ρ is well defined. The same argument shows that ρ is unique.

We call the triple $(\mathfrak{X}_C, f, \sigma)$ the <u>complex double</u> of $\mathfrak{X}$.

Lemma 1.6.3. Let $\mathfrak{X}$ be a (connected) Klein surface. X_C is disconnected if and only if X is orientable and $\partial X = \emptyset$.

Proof. Assume now that X is orientable and $\partial X = \emptyset$. Recall that in (1.2.4) we saw that by possibly using $-\overline{z}_j$ rather than z_j we can form an analytic atlas $\underset{\sim}{V} \equiv (U_j, w_j)_{j \in J} \in \mathfrak{X}$. Since $(\mathfrak{X}_C, f, \sigma)$ is unique, we may replace $\underset{\sim}{U}$ by $\underset{\sim}{V}$, so we assume that $\underset{\sim}{U}$ is analytic; then no identifications of the first type occur between U_j' and U_j''. Since $\partial X = \emptyset$, no identifications of the second type occur. As a result $\mathfrak{X}_C$ has two components: $\mathfrak{X}_C'$, the images of the U_j' 's; and $\mathfrak{X}_C''$, the images of the U_j'' 's. Conversely assume that X_C is disconnected. Since X is connected and since $f : \mathfrak{X}_C \longrightarrow \mathfrak{X}$ is an unramified double covering, X_C has two components Y and Z, each of which is mapped homeomorphically via f onto X. Since X_C is orientable and without boundary so is Y, thus so is X $(= f(Y))$.

There are two other useful unramified double covers of $\mathfrak{X}$, the <u>orienting double</u> $\mathfrak{X}_0$, and the classical or <u>Schottkey double</u> $\mathfrak{X}_S$ of $\mathfrak{X}$. They are both obtained via constructions similar to that employed above for the complex double.

We now investigate the relation between $E(\mathfrak{X})$ and $E(\mathfrak{X}_C)$. When

X is orientable and without boundary, then $E(\mathfrak{X}_C) \cong E(\mathfrak{X}) \times E(\mathfrak{X})$. Hence we restrict ourselves to the case where X_C is connected.

Since $f\sigma = f$, $\sigma^* f^* = f^*$, and $f^*(E(\mathfrak{X}))$ is contained in the subfield of $E(\mathfrak{X}_C)$ fixed by σ. Actually these two fields are equal.

Theorem 1.6.4. If X_C is connected then, $f^*(E(\mathfrak{X}))$ equals the fixed field of $E(\mathfrak{X}_C)$ under σ^*.

Proof. Let $\underline{g} \in E(\mathfrak{X}_C)$ satisfy $\sigma^*(\underline{g}) = \underline{g}$, and let (U,z) be a dianalytic chart on $\mathfrak{X}$. Then $g_{\tilde{U}}$ satisfies

$$g_{\tilde{U}} \cdot \sigma = \varkappa \cdot g_{\tilde{U}}$$

($\varkappa$ denotes complex conjugation). If $x \in U$, let x', x'' denote the corresponding points in U', U'', and define h_U by

$$h_U(x) = g_{\tilde{U}}(x') = \overline{g_{\tilde{U}}(x'')}.$$

It is easily checked that $\{h_U\}$ form a dianalytic function $\underline{h}$ on $\mathfrak{X}$ with $f^*\underline{h} = \underline{g}$.

Corollary 1.6.5. If X_C is connected, then $E(\mathfrak{X}_C) = (f^*E(\mathfrak{X}))(\sqrt{-1})$.

Corollary 1.6.6. There exist non-constant meromorphic functions on every Klein surface.

Proof. Use the corresponding result about Riemann surfaces. For the compact case see [G_2], for the non-compact case see [G_2R].

To construct $\mathfrak{X}_o$ proceed as above to construct Ω but employ only identifications of the first type and let X_o be the resulting identification space. Using essentially the same argument as employed above, except that identifications of the second type are not employed we have the following.

Theorem 1.6.7. There exists a double cover $f : \mathfrak{X}_o \longrightarrow \mathfrak{X}$ of $\mathfrak{X}$ by a Riemann surface with boundary $\mathfrak{X}_o$ such that

(i) $f^{-1}(\partial X) = \partial X_o$

(ii) $\mathfrak{X}_o$ has an antianalytic involution σ with $f\sigma = \sigma$. If $(\mathfrak{X}_o', f', \sigma')$ is any other such triple, then there is a unique analytic isomorphism $\rho : \mathfrak{X}_o' \longrightarrow \mathfrak{X}_o$ with $f' = f\rho$.

Further, f is unramified; σ is the only antianalytic automorphism of $\mathfrak{X}_o$ such that $f\sigma = f$; and X_o is disconnected if and only if X is orientable.

The proof of 1.6.7 is obtained by a straight-forward modification of the proofs of 1.6.1, 1.6.2 and 1.6.3. In particular, we have the following analogue of (1.6.2):

Proposition 1.6.8. If $\mathfrak{Y}$ is a Riemann surface with boundary and $h : \mathfrak{Y} \longrightarrow \mathfrak{X}$ is a morphism with $h^{-1}(\partial X) = \partial Y$, then there is a unique analytic morphism $g : \mathfrak{Y} \longrightarrow \mathfrak{X}_o$ such that the following diagram commutes:

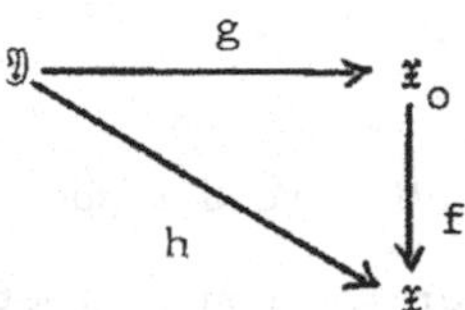

Example 1.6.3. If X is a Möbius strip X_o is an annulus and X_C a torus. If X is an annulus then X_o is a disjoint union of two annuli and X_C a torus.

Schottky applied his doubling procedure to orientable surfaces, however, it can be applied to non-orientable Klein surfaces in either of two ways. We could modify the above procedure so that identifications always occur between X_j' and X_k' (and X_j'' and X_k''). Or, more directly, one can take two copies of $\mathfrak{X}$, $\mathfrak{X}'$ and $\mathfrak{X}''$, with opposite orientations and glue them together on the boundary to form $\mathfrak{X}_S$. X_S then has some interesting properties. First of all, if X is orientable $\mathfrak{X}_S = \mathfrak{X}_C$. Secondly, if X is non-orientable then so is X_S. Of course, X_S is disconnected if and only if $\partial X = \emptyset$. Example: if X is a Möbius strip, X_S is a Klein bottle. (Since the Schottky double will not be used to prove the basic existence theorems on Klein surfaces, we have adopted a less formal style in dealing with it.)

Proposition 1.6.9. Let $\mathfrak{X}$ be a compact Klein surface and let $f : \mathfrak{Y} \longrightarrow \mathfrak{X}$ be an unramified double cover of $\mathfrak{X}$. Then the Euler characteristics satisfy $\chi(Y) = 2\chi(X)$.

Proof. Let T be a triangulation of $\mathfrak{X}$, with e_j the number of j-cells in T such that T lifts to a triangulation T' of Y, with e_j' the number of j-cells. Let B be a component of ∂X such that $f^{-1}(B)$ is mapped bijectively to B. Let b be the total number of vertices on such components B. Then

$$\begin{aligned} e_2' &= 2e_2 \\ e_1' &= 2e_1 - b \\ e_0' &= 2e_0 - b, \end{aligned}$$

and $\chi(Y) = 2\chi(X)$.

$\mathfrak{X}_C$, $\mathfrak{X}_O$, and $\mathfrak{X}_S$ are of course, not the only unramified double covers of $\mathfrak{X}$. However, if X is compact, there are (up to isomorphism) only finitely many unramified double covers of $\mathfrak{X}$. We can use the methods of §5 to determine all of them.

Proposition 1.6.10. Let $\mathfrak{X}$ be a compact Klein surface such that ∂X has $r \geq 1$ components. Then there are $2^a - 1$ (connected) unramified double covers of $\mathfrak{X}$, where $a = r + 1 - \chi(X)$.

Proof. Since $\partial X \neq \emptyset$, the fundamental group of X is free of rank $1 - \chi(X)$ and there are $2^{1-\chi(X)}$ topological double covers $g : Z \longrightarrow X$ of X, including the case where Z consists of two disjoint copies of X [M]. Using results in §5, each Z carries a unique dianalytic structure $\mathcal{B}$ making g a morphism. If B is a component of ∂X, then

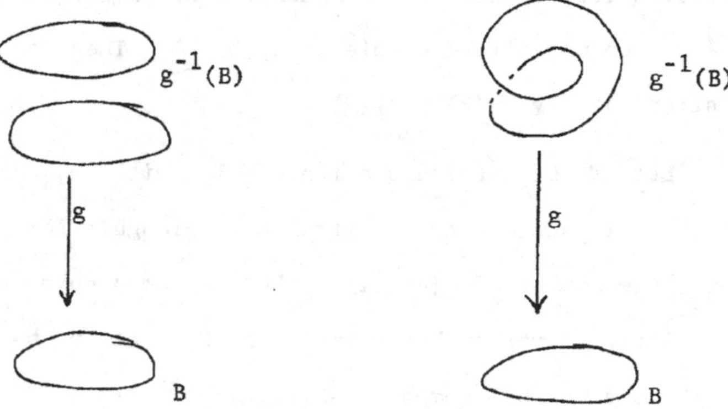

$g^{-1}(B)$ having two or one components. In either case we can use the identification procedure of §5 to pass in two steps to a situation where $g^{-1}(B)$ is a circle in the interior of Z, which is mapped homeomorphically to B: e.g., in the second case we have,

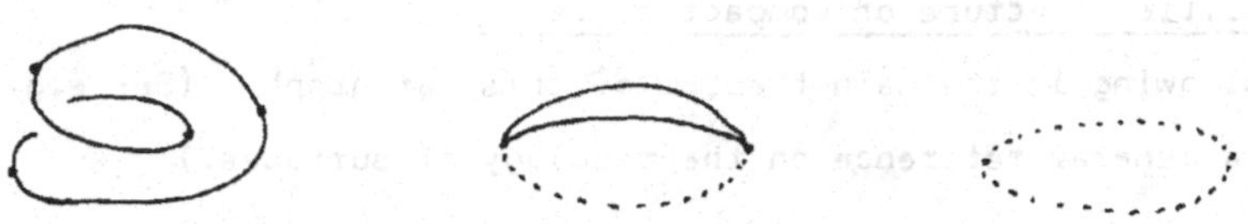

For each of the r components of ∂X we have the option of applying this identification procedure, and hence we have constructed $2^r\, 2^{1-\chi(X)}$ unramified double covers of $\mathfrak{X}$ (among them the union of two disjoint copies of $\mathfrak{X}$). Any unramified double cover of $\mathfrak{X}$ is easily seen to be isomorphic to one of these, since its restriction to $X - \partial X$ is a topological covering space.

§7. Dianalytic structure on compact surfaces

The following is the main theorem of this paragraph. (See e.g. [M] as a general reference on the topology of surfaces.)

Theorem 1.7.1. Let X be a (connected) compact surface. There exists a dianalytic structure $\mathfrak{X}$ on X; thus $\mathfrak{X}$ is a Klein surface.

Proof. Assume first that X is orientable and without boundary. X then is characterized by its Euler characteristic, $\chi(X)$. Recall that $\chi(X) = 2 - 2g$, g being a non-negative integer which is the genus of X. Let X' be a double covering of Σ, the sphere be ramified at $4 - \chi(X)$ points, f' being the covering map. Then $\chi(X') = \chi(X)$, and so X and X' are homeomorphic. Let them be identified. By (1.5.2), X has a dianalytic structure $\mathfrak{X}$ making $f : \mathfrak{X} \longrightarrow \Sigma$ a morphism of Klein surfaces. (Since X is orientable $\mathfrak{X}$ can be chosen to be analytic, such that f is analytic.)

Now assume that $\partial X \neq \emptyset$ or that X is non-orientable. We will represent X as a ramified covering of the closed disc D and apply (1.5.2) again. First assume that X is non-orientable and that ∂X has r components. Then $\chi(X) = 2 - r - q$, q being an integer, $q \geq 1$; further, X is characterized among all non-orientable surfaces by these integers r and q. Let X_0 be a double cover of D that is ramified at q points in $D - \partial D$. Then by (1.5.2) there is a dianalytic structure $\mathfrak{X}_0$ on X_0 making the covering map f_0 of X_0 onto D a morphism. Clearly X_0 is orientable and $\partial X_0 = f_0^{-1}(\partial D)$, ∂X_0 having one or two components according as q is odd or even. Assume first that $r = 0$. Choose two points on ∂D and let $I^{(1)}$ and $I^{(2)}$ be the two intervals in ∂D having these points as endpoints; thus $I^{(1)} \cup I^{(2)} = \partial X$. Apply the identification procedure in (1.5.4) to the components of

∂X_0 over $I^{(1)}$, and then to those over $I^{(2)}$. The resulting surface X' carries a dianalytic structure $\mathfrak{X}'$, by (1.5.4), making the induced covering map f' into a morphism. Then $\chi(X') = 2 - q$, as required. To show that X' is non-orientable it suffices to give a path on X' for which orientation is reversed as a result of making one traverse of the path. Such a path can be obtained by first looping around a ramified point x' of f' on X' and then passing across the fold over $(f')^{-1}(\partial D)$ and back to the original point. A section of $f'|X' \to D'$ with such a path is illustrated below.

(1.7.a)

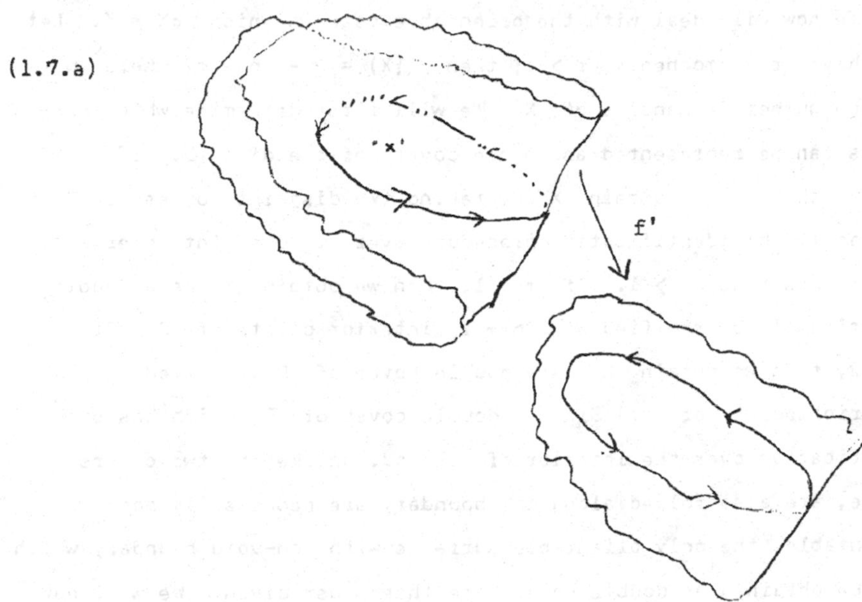

Hence X' is homeomorphic to X, and as a consequence we see that X can carry a dianalytic structure on it. In particular we see that the real projection plane $(q = 1)$ and the Klein bottle $(q = 2)$ can carry dianalytic structure.

Assume now that X is non-orientable and that ∂X has $r \geq 1$

components. Let $\mathfrak{X}_0$ be as it was above. Choose $2r$ points, $p_1,\dots,p_{2r}$ in order on ∂D. Let $I^{(j)}$ be the interval on ∂D from p_{2j-1} to p_{2j}. $f_0^{-1}(I^{(j)})$ consists of two disjoint intervals in ∂X_0 to which we apply the identification procedure of (1.5.4) to each of the r intervals in order and obtain $f' : \mathfrak{X}' \to \mathfrak{D}$, which by (1.5.4) is morphism. Clearly X' has r boundary components. On refering to (1.7.a) again it is obvious that X' is non-orientable. Since the Euler characteristics of X and X' are equal, these two surfaces are homeomorphic.

We now will deal with the orientable case in which $\partial X \neq \emptyset$. Let ∂X have r components $r \geq 1$; then $\chi(X) = 2 - 2p - r$, where p is the number of handles of X. We will first determine which surfaces can be represented as double covers of the disc D. If $p = 0$, then we can obtain X by taking two disjoint copies of D and apply the identification procedure over r disjoint intervals. Assume now that $p \geq 1$. If $r = 1$, then we obtain X as a double covering of D, ramified at $2p + 1$ interior points of D. If $r = 2$, then we obtain X as a double cover of D, ramified at $2p$ interior points of D. Since a double cover of D which has both ramification over the interior of D and, unlike the two covers above, are also folded along the boundary are necessarily non-orientable, the only orientable surfaces with non-void boundary which may be obtained as double covers are those just given. We will now show that if $r \geq 3$ and $p \geq 1$, can always be realized as triple coverings of D. Let X_0 be a double cover of D, ramified at $2p + 2$ interior points of D, and let D_0 be another copy of D. Let $I^{(1)},\dots,I^{(r-1)}$ be disjoint intervals in ∂D. We now identify D_0 and X_0 over $I^{(1)},\dots,I^{(r-1)}$, in order, for each choosing the interval in ∂X_0 from the same component. (Since $2p + 2$ is even

∂X_0 has two components.) Consider the following illustration.

(1.7.b)

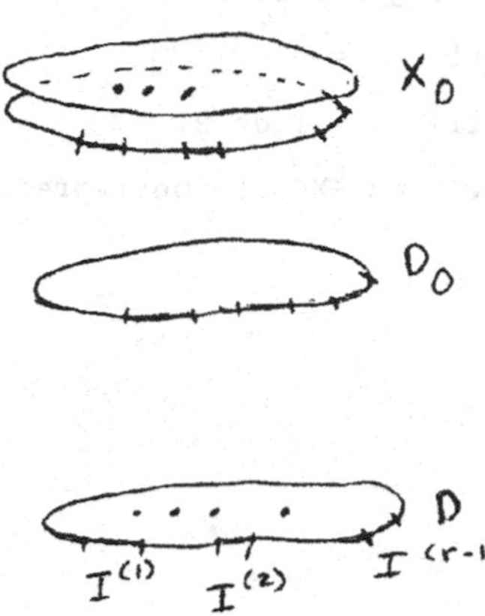

It is clear that the surface X' obtained in this way is orientable, and has r boundary components. Let us now compute its Euler characteristic. $\chi(X_0 \cup D_0) = \chi(X_0) + \chi(D_0) = (2 - (2p + 2)) + 1$. Each identification over an $I^{(j)}$ lowers the Euler characteristic by one, so $\chi(X') = \chi(X_0 \cup D_0) - (r - 1) = 2 - 2p - r$, as required. On appealing to (1.5.2) and (1.5.4) the theorem is proved.

The proof above gives us more than asserted, namely the following.

Theorem 1.7.2. Let X be a compact (connected) surface.

1) If X is orientable and if $\partial X = \emptyset$, then it can be represented as a double cover of the Riemann sphere.

2) If X is non-orientable, then it can be represented as a double cover of the disc.

3) If X is orientable and $\chi(X) = 2 - 2p - r$, where ∂X has r components, then X can be represented as a double cover of the disc only if the following occur:

(i) $p = 0$

(ii) $r = 1$ or 2.

If $p \geq 1$ and $r \geq 3$, then X can be represented as a triple cover of the disc.

§8. Quotients of Klein Surfaces

In this section we will show that the quotient space of a Klein surface under a discontinuous group of automorphisms has a canonical Klein surface structure. We will start with some results on involutions.

Let A be the group of all germs of dianalytic automorphisms of neighborhoods of zero in C. The elements of A posess convergent power series either in z or in $\bar{z}$ in some neighborhood of 0 in C, whose constant term is non-zero. Let ι denote the identity of A and let κ denote germ of the map $z \to \bar{z}$ in A.

Proposition 1.8.1. Let $\sigma \in A$ be an antianalytic involution; then σ is conjugate to κ in A (i.e., there exists $\beta \in A$ such that $\sigma = \beta^{-1}\kappa\beta$.)

Proof. Let $\Sigma_{n=1}^{\infty} a_n \bar{z}_n$ be the power series development for σ at zero. Since σ is an invalution (i.e., $\sigma^2 = \iota$), $|a_1| = 1$. Assume first that $a_1 \neq -1$. Let $\alpha = \kappa\sigma + \iota \in A$; then $\alpha\sigma = \kappa + \sigma$, proving that $\sigma = \alpha^{-1}\kappa\alpha$. If $a_1 \neq 1$, let $\delta = \kappa\sigma - \iota \in A$ and note that $\delta\sigma = (-\kappa)\delta$, showing that σ and $-\kappa$ are conjugate. Since $-\kappa = (-i\iota)\kappa(i\iota)$, we see that $-\kappa$ and κ are conjugate. Hence σ and κ are conjugate, proving the proposition.

Let A^+ be the group of all germs of dianalytic automorphisms of neighborhoods of zero in C^+. Each element σ of A^+ admits a power series development in either z or $\bar{z}$. Such series have zero constant term, non-zero linear term, and real coefficients; thus A^+ is a subgroup of A. Further if σ is analytic (resp. antianalytic) the linear term in its power series is positive (resp. negative).

Proposition 1.8.2. Let $\sigma \in A^+$ be an antianalytic involution.

σ is conjugate to $-\mathcal{K}$ in A^+.

Proof. $\sigma = \Sigma_{n=1}^{\infty} a_n \bar{z}^n$. Since σ is an involution $|a_1| = 1$. Since $\sigma \in A^+$, a_1 is real; thus $a_1 = \pm 1$. Since σ is antianalytic, $a_1 = -1$, as remarked above. As noted in the proof of (1.8.1) above, σ is conjugate to $-\mathcal{K}$ under $\mathcal{K}\sigma - \iota$, proving the proposition.

Proposition 1.8.3. Let $\theta \in A^+$ be an element of finite order. Then either θ is an antianalytic involution or θ is the identity map.

Proof. θ^2 is analytic and of finite order; this it suffices to show that the only analytic element β in A^+ of finite order is ι. Indeed, let $\beta(z) = \Sigma_{n=1}^{\infty} b_n z^n$. Since β is of finite order k, $|b_1| = 1$. Since b_1 is necessarily positive, $b_1 = 1$. Assume, for the moment, that $\beta \neq \iota$: i.e., assume that $k > 1$. Since $\beta^k = \iota$, $k\beta^{k-1}(z)\beta'(z) = 1$ for all z in some neighborhood N of zero in C. In particular $k\beta^{k-1}(0)\beta'(0) = 1$. Since $b_1 = 1$, $\beta^{k-1}(0) = 1/k$; but $\beta^{k-1} \in A^+$ and hence its value at zero is zero, which is absurd, proving that $\beta = \iota$ and hence proving the proposition.

A group G acts <u>discontinuously</u> on a space X if for each $x \in X$ there is a neighborhood U of x such that $\theta(U) \cap U$ is empty, for all but a finite number of $\theta \in G$. An autohomeomorphism of a Klein surface that is dianalytic will be called an <u>automorphism</u>.

Theorem 1.8.4. Let $\mathfrak{X}$ be a Klein surface and let G be a group of automorphisms of $\mathfrak{X}$, which act discontinuously on X. Then the quotient space $Y = X/G$ has a unique dianalytic structure $\mathfrak{Y}$ such that the canonical map π of $\mathfrak{X}$ onto $\mathfrak{Y}$ is a morphism of Klein surfaces.

Proof. The uniqueness of the dianalytic structure on X/G follows from (1.4.5). It is well known that X/G is a Hausdorff space (see e.g. [SS]). We will now put a dianalytic structure on X/G; thus it is a surface with boundary.

Let $x \in X$, and let $S_x = \{\theta \in G : \theta(x) = x\}$. ($S_x$ is necessarily a finite subgroup of G.) We can find a dianalytic chart (U,z) such that $x \in U$, $z(x) = 0$, U is connected, $\theta(U) = U$ for all $\theta \in S_x$, and $\theta(U) \cap U = \emptyset$ for all $\theta \in G - S_x$. (See e.g., [SS] for details.) Let $S'_x = \{\theta \in S_x : z\cdot\theta\cdot z^{-1}$ is analytic$\}$. Clearly S'_x is independent of the choice of (U,z). Either $S_x = S'_x$, or S'_x is a subgroup of S_x of index two. We may choose U sufficiently small to ensure that x is the only element of U which is left fixed by the non-identity elements of S'_x.

First consider the case in which $S_x = S'_x$. Then $V = \pi(U) \simeq U/S_x \simeq U/S'_x$, where here S_x is regarded as maps restricted to U. The function $\prod_{\theta\in S'_x} z\cdot\theta$ is invariant under S'_x, and hence induces a function w on V for which $w\cdot\pi = \prod_{\theta\in S'_x} z\cdot\theta$. If we let $g = \prod_{\theta\in S'_x} z\cdot\theta\cdot z^{-1}$, then g has a zero of order n (the order of S'_x) at the origin, and is analytic on $z(U)$. Possibly by restricting U still further, we may conclude that g takes no value more than n times on $z(U)$. Since $\pi|U$ is n-to-1, except at x, we deduce from the following commutative diagram,

(1.8.1) 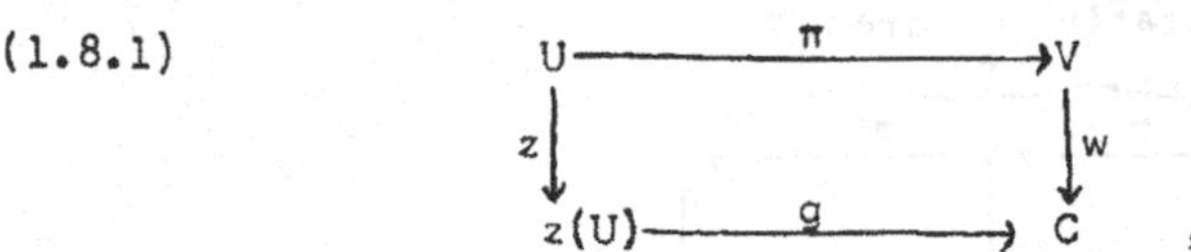

that w is one-to-one. If $x \in \partial X$, then by (1.8.3), S'_x has order 1; thus $\pi|U$ is a bijection, g is the identity map, and w is a homeomorphism onto an open subset $w(V) = z(U)$ in C^+. Assume now that $x \notin \partial X$. Then $z(U)$ is an open subset in C. Let D be an open subset of V. Since g and z are open mappings, $w(D) = g \cdot z \cdot (\pi^{-1}(D))$ is open in C.

We now consider the case in which $S_x \neq S'_x$ and $x \notin \partial X$. Let $V' = U/S'_x$ and let $w' : V' \to C$ as defined above. $I_x = S_x/S'_x$ is a group of order two which acts on V'. Let σ generate I_x. $w' \cdot \sigma \cdot (w')^{-1}$ is an antianalytic invalution of the neighborhood $w'(V')$ of zero in C. Using (1.8.1), and possibly shrinking U, we find an analytic function h on $w'(V')$ with $h \cdot w' \cdot \sigma \cdot (w')^{-1} \cdot h^{-1}$ equal to complex conjugation. $\pi|U$ factors as follows:

(1.8.2) $U \xrightarrow{\pi} V$, $\quad U \xrightarrow{\pi'} V' \xrightarrow{\pi''} V$

The function $\varphi \cdot h \cdot w'$ (φ being the folding map) is invariant under σ. This may be seen by consulting the following commutative diagram:

(1.8.3)

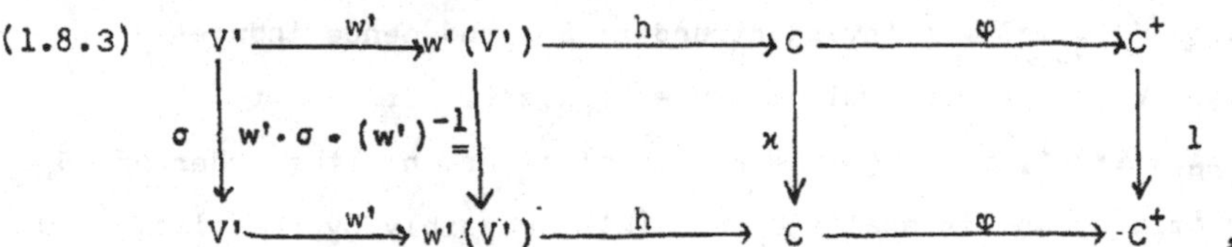

where $\varkappa$ denotes complex conjugation. Hence $\varphi \cdot h \cdot w'$ induces a function w on $V = \pi(U) \simeq V'/I_x$, so that $w \cdot \pi' = \varphi \cdot h \cdot w'$. We then have the following commutative diagram.

(1.8.4)

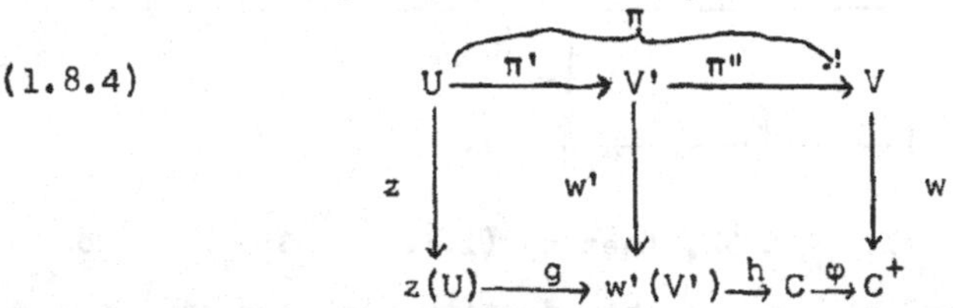

It is easily checked that w is a bijection and is a homeomorphism onto an open image in C^+.

The last case to consider is $S_x \neq S_x'$ and $x \in \partial X$. By Proposition 1.8.3, S_x' is the trivial group, and as a result $S_x = I_x$ is of order two. Let σ generate S_x. Then $z \circ \sigma \circ z^{-1}$ is an antianalytic involution of a neighborhood of zero in C^+. Shrinking U if necessary and using Proposition 1.8.2, we can find an analytic homeomorphism f, defined on z(U), such that $(-\bar{\lambda}) \circ f \circ z = f \circ z \circ \sigma$. Let $s(\lambda) = \lambda^2$ for all $\lambda \in C$. We then have the following commutative diagram:

(1.8.5)

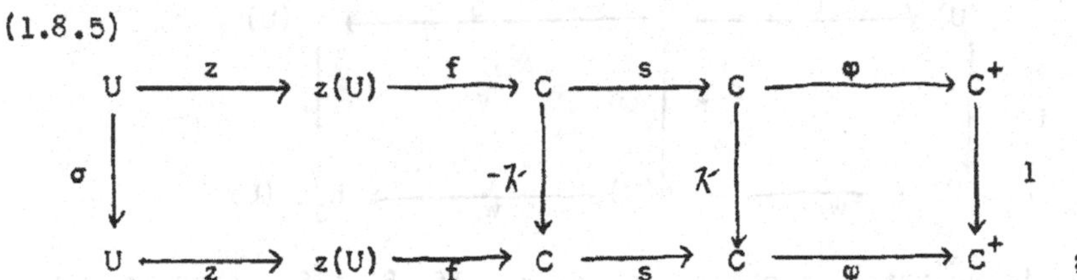

thus $\varphi \circ s \circ f \circ z$ is invariant under σ, and hence induces $w: V \longrightarrow C^+$, making the following diagram commutative:

(1.8.6)

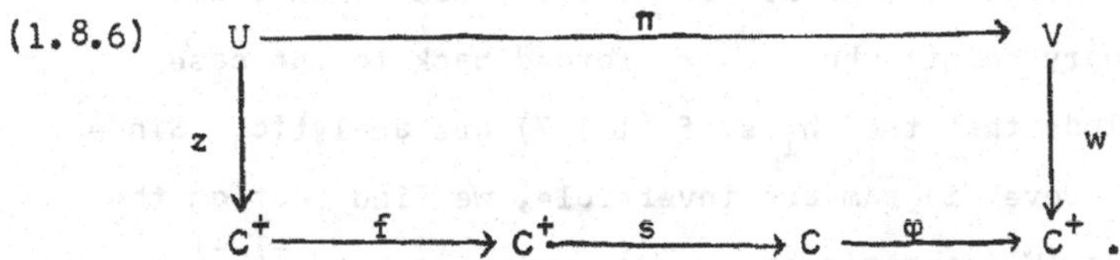

One can again easily see that w is a homeomorphism with an open image in C^+. (Hence we have shown that X/G is a surface, possibly with boundary.)

We now will show that the atlas $\{(V,w)\}$, just described, on X/G is dianalytic. Let the <u>index</u> of G at $y = \pi(x)$ be defined to be the order of S_x'. Note that for any V there is at most one point $y \in V$ such that the index of G at g is greater than one.

Hence, we may pass to a subatlas $\{(V_i,w_i)\}$ such that if $i \neq j$, $V_i \cap V_j$ contains no point of index greater than one. Let (V_i,w_i) come from a chart (U_i,z_i) on X. Now let $i \neq j$ and consider the function $w_i w_j^{-1}$ on $w_j(V_i \cap V_j)$. On utilizing the Schwarz reflection principal, we may restrict our attention to interior points. Let y be an interior point of $V_i \cap V_j$, let x be a pre-image of y under π in U_i, and let U be an open connected neighborhood of x in $U_i \cap \pi^{-1}(V_i \cap V_j)$ such that $\pi|U$ is injective. We then have the following commutative diagram:

(1.8.7)

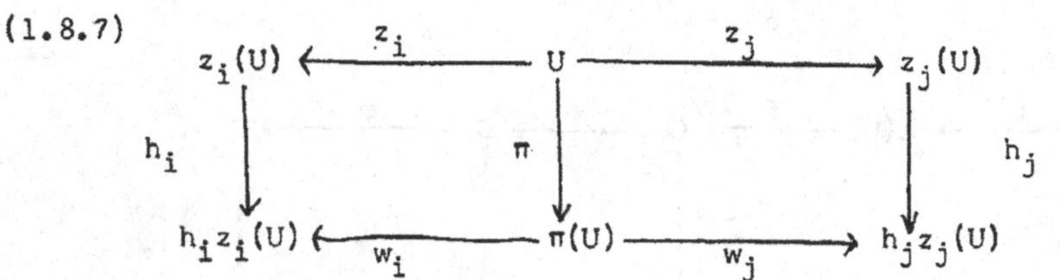

Since y is an interior point, so is x. If $S_x' = S_x$, then we are in the first case treated, with $n = 1$. The h's then are the g's refered to in Diagram (1.8.1), and are analytic. Were the case in which $S_x' \neq S_x$ to occur, then Diagram (1.8.4) would hold, and y would be a boundary point; thus we are forced back to the case above, and conclude that the h_i's of (1.8.7) are analytic. Since all maps in the above diagram are invertible, we find that on the set $w_j\pi(U) = h_j z_j(U)$, containing $w_j(y)$, $w_i w_j^{-1} = h_i z_i z_j^{-1} h_j^{-1}$. Hence the family $\{(V_i,w_i)\}$ makes X/G into a Klein surface. Diagrams (1.8.1), (1.8.4), and (1.8.6) taken together with the Theorem 1.5.2 prove that π is a morphism, proving the theorem.

§9. Surfaces of genus 0 and 1

In this section we will determine all dianalytic structures on the disc, real projective plane, annulus, Möbius strip, and Klein bottle. If $\mathfrak{X}$ is a Klein surface, X being one of the above, then X_C is either a sphere (in the first two cases), or a torus. It is well known that the sphere carries a unique analytic structure on it (up to isomorphism), and that the distinct analytic structures on the torus are parametrized by the fundamental domain of the modular group on $\mathbb{C}^+$.

Proposition 1.9.1. Let $\mathfrak{Y}$ be a Riemann surface, and let σ be an anti-analytic involution of $\mathfrak{Y}$. Then $(\mathfrak{Y}/\sigma)_C$ is canonically isomorphic to $\mathfrak{Y}$.

Proof. From the universal property of the complex double (1.6.2), we have an analytic map $\rho : \mathfrak{Y} \longrightarrow (\mathfrak{Y}/\sigma)_C$ which makes the following diagram commute:

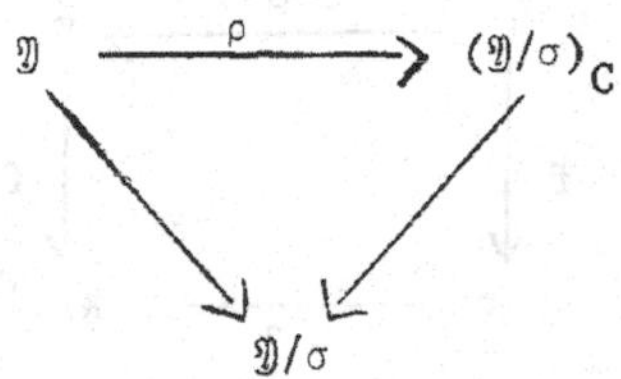

It is easily checked that $(\mathfrak{Y}/\sigma)_C$ is connected, and hence that ρ is an isomorphism.

Proposition 1.9.2. Let $\mathfrak{Y}$ be a Riemann surface, and let σ, σ' be anti-analytic involutions of $\mathfrak{Y}$. Then the Klein surfaces $\mathfrak{X} = \mathfrak{Y}/\sigma$ and $\mathfrak{X}' = \mathfrak{Y}/\sigma'$ are isomorphic if and only if there is an automorphism θ of $\mathfrak{Y}$ such that $\sigma' = \theta\sigma\theta^{-1}$.

Proof. Let $\eta : \mathfrak{X} \longrightarrow \mathfrak{X}'$ be such an isomorphism. Then, using (1.9.1) and (1.6.2), we obtain θ and the following commutative diagram:

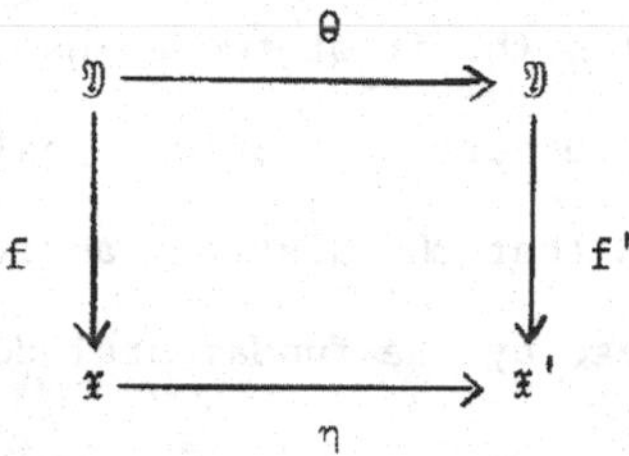

Since $\theta\sigma\theta^{-1}$ is an anti-analytic involution of $\mathfrak{Y}$, and since

$$f'\theta\sigma\theta^{-1} = \eta f\sigma\theta^{-1} = \eta f\theta^{-1}$$
$$= f'\theta\theta^{-1} = f';$$

then $\theta\sigma\theta^{-1} = \sigma'$.

Conversely, assume that such an automorphism θ of $\mathfrak{Y}$ exists. Since $f'\theta\sigma = f'\sigma'\theta = f'\theta$, the map $f'\theta$ factors through f:

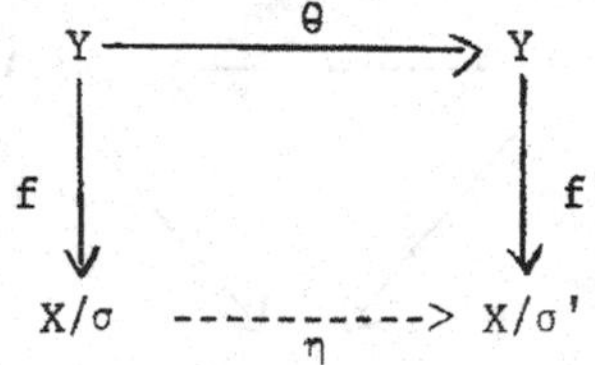

The map η, which is a priori only continuous, is a morphism, by (1.4.3), and is easily seen to be bijective.

Corollary 1.9.3. Let $\mathfrak{X}, \mathfrak{X}'$ be Klein surfaces. Then $\mathfrak{X}$ and $\mathfrak{X}'$ are isomorphic if and only if there is an isomorphism $\theta : X_C \longrightarrow X'_C$ for which $\theta\sigma\theta^{-1} = \sigma'$.

We will now show that both the disc and the real projective plane have unique dianalytic structures. The disc is obtained from Σ via the anti-analytic involution $\sigma_1(z) = \bar{z}$, while the projective plane comes from $\sigma_2(z) = -1/\bar{z}$.

Theorem 1.9.4. Every anti-analytic involution of Σ is conjugate to either σ_1 or σ_2.

Proof. Let σ be an anti-analytic involution of Σ. Then $\sigma \kappa$ is an analytic automorphism of Σ, and hence is a linear fractional transformation. Hence

$$\sigma(z) = \frac{a\bar{z} + b}{c\bar{z} + d}$$

and we may assume that

$$\det \begin{pmatrix} a & b \\ c & d \end{pmatrix} = \det A = 1.$$

Since σ is an involution, $\bar{A}A = \pm I$. If θ is an analytic automorphism of Σ represented by a matrix B, then $\theta^{-1}\sigma\theta$ is represented by $\bar{B}AB^{-1}$. Hence we may regard A as the matrix of a conjugate linear operator on $\mathbb{C}^2$, where conjugation by θ corresponds to the choice of a new basis in $\mathbb{C}^2$.

Case I: $\bar{A}A = -I$. Let ψ be the conjugate linear operator on $\mathbb{C}^2$ whose matrix is A. If α were an eigenvector of ψ, say $\psi(\alpha) = \lambda\cdot\alpha$, λ being a scalar, then for all such $\alpha \in C^2$.

$$\begin{aligned} -\alpha = \psi^2(\alpha) &= \psi(\lambda\cdot\alpha) = \bar{\lambda}\,\psi(\alpha) \\ &= \bar{\lambda}\,\lambda\cdot\alpha, \end{aligned}$$

which is impossible. Hence if $0 \neq \alpha \in C^2$, then $\{\alpha, \psi(\alpha)\}$ is a basis for $\mathbb{C}^2$, and the matrix of ψ with respect to this basis is

$$\begin{pmatrix} 0 & -1 \\ 1 & 0 \end{pmatrix}.$$

Hence σ is conjugate to σ_2.

Case II: $\bar{A}A = I$. If every vector in $\mathbb{C}^2$ were an eigenvector of the conjugate linear operator ψ, then ψ would be a scalar multiple of the identity map, and this is not conjugate linear (note that ψ can be regarded as a linear operator on $\mathbb{R}^4$). Let α be a non-eigenvector of ψ. Then $\{\alpha, \psi(\alpha)\}$ is a basis for $\mathbb{C}^2$, and hence so is

$$\{\alpha + \psi(\alpha),\ i\alpha - i\psi(\alpha)\}.$$

But the matrix of ψ, with respect to the latter basis, is

$$\begin{pmatrix} 1 & 0 \\ 0 & 1 \end{pmatrix},$$

and σ is conjugate to σ_1.

Corollary 1.9.5. The disc and the real projective plane carry (up to isomorphism) a unique dianalytic structure.

<u>Remark</u>. We will be able to give another proof of this fact in the second chapter. The disc and the real projective plane are the two algebraic curves of genus zero whose constant field is $\mathbb{R}$. The former corresponds to $R(x)$ and the latter to $R(x,y)$, where $x^2 + y^2 = -1$. The second of these fields frequently is used to give an example of a real curve without a real point.

We shall now classify all Klein surfaces $\mathfrak{X}$ such that X_C is a torus. Since the Euler characteristic of X_C is zero, and is twice the Euler characteristic of X, we see that then X must be either an annulus, a Möbius strip, or a Klein bottle. These are easily distinguished by having 2, 1, and 0 boundary components, respectively.

First we recall briefly the classification of analytic structures on the torus (see [G_2]). If $\mathfrak{D}$ is a torus with analytic structure (a compact Riemann surface of genus 1), then the universal covering space of $\mathfrak{D}$ is isomorphic to $\mathbb{C}/\Lambda$, where Λ is a lattice in $\mathbb{C}$ (i.e., a discrete rank two subgroup). Λ may be normalized so that it is generated by $\{1,\gamma\}$ where γ satisfies:

(1.9.1)
$$-1/2 < \mathrm{Re}(\gamma) \leq 1/2,$$
$$|\gamma| \geq 1 \quad \text{when} \quad \mathrm{Re}\,\gamma \geq 0,$$
$$|\gamma| > 1 \quad \text{when} \quad \mathrm{Re}\,\gamma < 0.$$

Thus the analytic structures on a torus are parametrized by the points of the region (1.9.1), (which is just a fundamental region of the modular group, $GL(2,\mathbb{Z})$, on the upper half plane).

Let $\mathfrak{X}$ be such a Klein surface and let σ be the canonical anti-analytic involution of $\mathfrak{X}_C \cong \mathbb{C}/\Lambda$. Then σ lifts to an anti-analytic automorphism $\tilde{\sigma}$ of $\mathbb{C}$. This must have the form

$$\tilde{\sigma}(z) = a\bar{z} + b$$

$a \neq 0$, and b can be chosen to lie in the region

(1.9.2)
$$\{r + s\gamma \mid 0 \leq r, \; s < 1\},$$

a fundamental parallelogram.

Since $z \equiv z' \pmod{\Lambda}$ implies $\tilde{\sigma}(z) \equiv \tilde{\sigma}(z') \pmod{\Lambda}$, we must have

$$a \in \Lambda \tag{1.9.3}$$

$$a\gamma \in \Lambda. \tag{1.9.4}$$

Since $\sigma^2 = 1_{X_C}$, then from $\tilde{\sigma}^2(z) = a\bar{a}z + a\bar{b} + b$ we obtain

$$|a| = 1 \tag{1.9.5}$$

$$a\bar{b} + b \in \Lambda. \tag{1.9.6}$$

Proposition 1.9.6. If $\mathfrak{X}_C \cong \mathbb{C}/\Lambda$, with Λ satisfying (1.9.1), then if $|\gamma| > 1$, $\operatorname{Re}(\gamma) = 0$ or $\operatorname{Re}(\gamma) = 1/2$.

Proof. Assume $|\gamma| > 1$. Then (1.9.3) and (1.9.5),

$$a = \pm 1.$$

and then from (1.9.4), $\bar{\gamma} \in \Lambda$, and thus $\gamma + \bar{\gamma} \in \Lambda$. This is possible if and only if $\gamma + \bar{\gamma} \in \mathbb{Z}$, so $\operatorname{Re}(\gamma) = 0$ or $1/2$.

There are now five cases which we must consider, three general cases and two special ones with complex multiplication. We list the cases below, along with the possible values of a, satisfying (1.9.3), (1.9.4), and (1.9.5).

Case 1. $\operatorname{Re}(\gamma) = 0$, $|\gamma| > 1$, $a = \pm 1$.

Case 2. $\gamma = i$, $a = \pm 1, \pm i$.

Case 3. $|\gamma| = 1$, $0 < \operatorname{Re}\gamma < 1/2$, $a = \pm\gamma$.

Case 4. $\gamma = \omega = e^{2\pi i/6}$, $a = \pm 1, \pm\omega, \pm\omega^2$.

Case 5. $\operatorname{Re}(\gamma) = 1/2$, $|\gamma| > 1$, $a = \pm 1$.

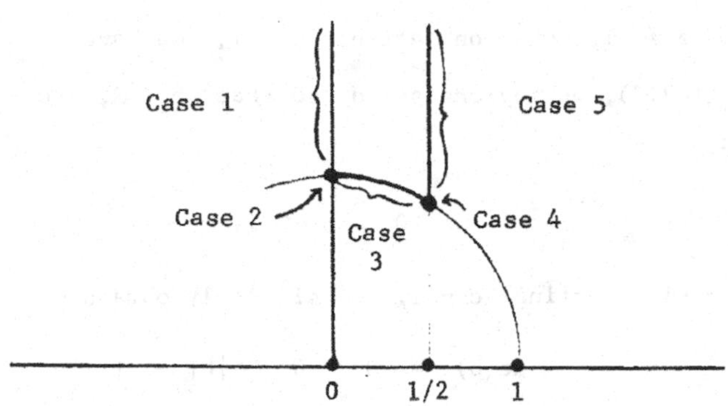

In order to classify Klein surfaces, we must, by (1.9.2), classify equivalence classes -- under conjugation -- of anti-analytic involutions. Let τ be an automorphis of $\mathfrak{T}_C$. Then τ lifts to an automorphism $\tilde{\tau}$ of C, which must have the form $\tilde{\tau}(z) = cz + d$. $\tilde{\tau}$ must take the fundamental domain (1.9.2) into another; thus it must be area-preserving. Hence d is arbitrary, while c must be ± 1 in the three general cases: i.e., in Cases 1, 3, and 5. In Case 2, c can also take the values $\pm i$, while in Case 4 it can take on the values $\pm\omega$, $\pm\omega^2$ in addition to ± 1. Thus in the general cases c is a square root of 1, in Case 2, a fourth root of 1, and in Case 4 a sixth root of 1. It is easily seen that

(1.9.7) $$\tilde{\tau}\tilde{\sigma}\tilde{\tau}^{-1}(z) = c^2 a \bar{z} + cb + (d - a c^2 \bar{d}), \text{ since } c/\bar{c} = c^2.$$

We now need the following simple lemma.

Lemma 1.9.7. If $\lambda \in C$, $|\lambda| = 1$, then

$$\{u + \lambda\bar{u} \mid u \in C\} = \{r\lambda^{1/2} \mid r \in \mathbb{R}\}.$$

Proof. It is easily checked that $(u + \lambda\bar{u})/\lambda^{1/2}$ is real, and the lemma follows.

When $a \neq -1$, then on setting $c = 1$, we have $(-a\,c^2)^{1/2} \notin \mathbb{R}$. Hence by (1.9.7), we may choose d so that $b \in \mathbb{R}$, so -- by (1.9.2) -- we have

(1.9.8) $$0 < b < 1.$$

When $a = -1$, setting $c = 1$, we similarly obtain

(1.9.9) $$\mathrm{Re}(b) = 0, \quad 0 \le |b| < |\gamma|.$$

In the three general cases $c^2 = 1$, so we see from (1.9.7) that distinct values of a correspond to non-conjugate involutions. In Case 2, setting $c = i$, we see from (1.9.7) that we may restrict ourselves to the cases

(1.9.10) $$a = 1, i,$$

with $a = -1 \sim a = 1$ and $a = -i \sim a = i$. In Case 4 we see in the same way that $a = 1 \sim a = \omega^2 \sim a = -\omega$, and that $a = -1 \sim a = \omega \sim a = -\omega^2$, so that we may take

(1.9.11) $$a = 1, \omega.$$

Combining (1.9.6), (1.9.8), (1.9.9), (1.9.10), and (1.9.11), we obtain the following result:

Theorem 1.9.8. The following is a complete list of anti-analytic involutions of compact Riemann surfaces of genus 1, one from each conjugacy class. The last column classifies the resulting surface X as: A -- an annulus; K -- a Klein bottle; and M -- a Möbius strip.

Case	a	b	Family
1	1	0	A
	-1	0	A
	1	1/2	K
	-1	$\gamma/2$	K
2	$1 \sim -1$	0	A
	$i \sim -i$	0	M
	$1 \sim -1$	1/2	K
3	γ	0	M
	$-\gamma$	0	M
4	$1 \sim -\omega$	0	M
	$\omega \sim -1$	0	M
5	1	0	M
	-1	0	M

The determination of the topological type of each surface is obtained by a simple analysis of the fixed point set of the corresponding involution. Note that each of the families has a natural connected one - parameter structure. The situation is illustrated in the following diagram which shows how each of the families A, K, and M is located with respect to the corresponding values of γ (which are here put in a straight line in order to make the diagram two-dimensional).

Remark. Note that annuli are paired with Klein bottles while Möbius strips stand alone. This situation reflects the fact that the Galois cohomology of an annulus is isomorphic to $\mathbb{Z}_2$, while that of a Möbius strip is trivial. Each Klein bottle is a principal homogeneous space for the corresponding annulus.

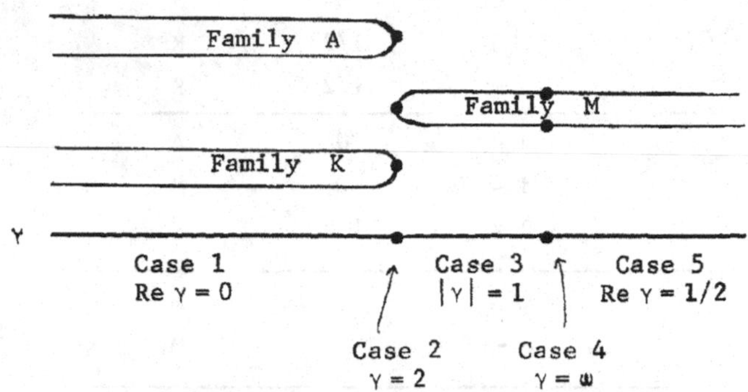

§10. Meromorphic differentials

Throughout this section let $D \equiv \partial/\partial z$ and let $\bar{D} \equiv \partial/\partial\bar{z}$. As before let $\varkappa$ denote complex conjugation. Let G be analytic and H anti-analytic; then note that the following hold:

*) $D(\varkappa \circ G \circ \varkappa) = \varkappa \circ D(G) \circ \varkappa$,

**) $D(\varkappa \circ H) = \varkappa \circ \bar{D}(H)$, and

***) $D(H \circ \varkappa) = \bar{D}(H) \circ \varkappa$.

Let $\mathfrak{X}$ be a Klein surface, let $\underset{\sim}{U} \equiv (U_j, z_j)_{j\in J} \in \mathfrak{X}$, and let T_{jk} be defined on each component V of $U_j \cap U_k$ as follows:

(1.10.1) $$T_{jk} \equiv \begin{cases} D(z_j \circ z_k^{-1}) \circ z_k, & \text{if } z_j \circ z_k^{-1} \text{ is analytic} \\ \bar{D}(z_j \circ z_k^{-1}) \circ z_k, & \text{if } z_j \circ z_k^{-1} \text{ is anti-analytic} \end{cases}$$

By a meromorphic differential on $\mathfrak{X}$ with respect to $\underset{\sim}{U}$ will be meant a family $\omega_{\underset{\sim}{U}} = (\omega_j)_{j\in J}$ of functions, $\omega_j : U_j \longrightarrow \Sigma$ such that the following hold:

a) $\omega_j \circ z_j^{-1}$ is meromorphic;

b) $\omega_j(U_j \cap \partial X) \subset R \cup \{\infty\}$;

c) on each component V of $U_j \cap U_k$,

$$\omega_k = \begin{cases} \omega_j\, T_{jk}, & \text{when } z_j \circ z_k^{-1} \text{ is analytic on } z_k(V) \\ \overline{\omega_j\, T_{jk}}, & \text{when } z_j \circ z_k^{-1} \text{ is anti-analytic on } z_k(V). \end{cases}$$

Such differentials can be added componentwise. Under this operation they form a real vector space. In order to free the notion of a meromorphic differential from the choice of $\underset{\sim}{U} \in \mathfrak{X}$, let $\eta_{\underset{\sim}{V}}$ be a meromorphic differential on $\mathfrak{X}$ relative to $\underset{\sim}{V} \in \mathfrak{X}$. We say that $\omega_{\underset{\sim}{U}}$ and $\eta_{\underset{\sim}{V}}$ are <u>equivalent</u> if $\omega_{\underset{\sim}{U}} \cup \eta_{\underset{\sim}{V}}$ is a meromorphic differential on $\mathfrak{X}$ relative to $\underset{\sim}{U} \cup \underset{\sim}{V}$, and we write $\omega_{\underset{\sim}{U}} \sim \eta_{\underset{\sim}{V}}$. It is clear that this relation is reflexive and symmetric. We leave to the reader the tedious and straightforward verification that it is transitive, using the chain rule and *), **), ***). It is also easily checked that if $\omega_{\underset{\sim}{U}}$ is a meromorphic differential and $\underset{\sim}{V} \in \mathfrak{X}$, then there is a unique meromorphic differential $\eta_{\underset{\sim}{V}}$ such that $\omega_{\underset{\sim}{U}} \sim \eta_{\underset{\sim}{V}}$.

An equivalence class under this relation will be called a <u>meromorphic differential</u> on $\mathfrak{X}$. Since $\varkappa$ is an $\mathbb{R}$- linear map, the set $\Omega(\mathfrak{X})$ of all such differentials is a vector space over $\mathbb{R}$.

If we take for $\underset{\sim}{U}$ the maximal dianalytic atlas on $\mathfrak{X}$, then we see that to each dianalytic chart (U,z) there corresponds a function ω_U on U (ω_U depends, of course, on U and on z).

Given $\underline{f} \in E(\mathfrak{X})$ and $\underset{\sim}{U} = (U_j, z_j) \in \mathfrak{X}$, we use the family $(f_j) = f_{\underset{\sim}{U}} \in \underline{f}$ to define a meromorphic differential $df_{\underset{\sim}{U}} = (df_j)$, where $df_j = D(f_j \circ z_j^{-1}) \cdot z_j$. It is easily verifed that this does define a meromorphic differential with respect to $\underset{\sim}{U}$, and that $df_{\underset{\sim}{U}} \sim df_{\underset{\sim}{V}}$. We denote the equivalence class of $df_{\underset{\sim}{U}}$ by $d\underline{f}$.

If $\underline{f} \in E(\mathfrak{X})$, $\omega \in \Omega(\mathfrak{X})$, and $\underset{\sim}{U} = (U_j, z_j) \in \mathfrak{X}$, we define a meromorphic differential with respect to $\underset{\sim}{U}$ by the family $(f_j\ \omega_j)$. It is easily checked that this is independent of $\underset{\sim}{U}$ and defines a meromorphic differential $\underline{f}\omega$ on $\mathfrak{X}$.

Lemma 1.10.1. $\Omega(\mathfrak{X})$ is a vector space over $E(\mathfrak{X})$. Further $d: E(\mathfrak{X}) \longrightarrow \Omega(\mathfrak{X})$ is a derivation. Finally, $d\underline{f} = 0$ if and only if $\underline{f} \in E_0(\mathfrak{X})$, the field of constant meromorphic functions on $\mathfrak{X}$.

The statements above can be checked locally, where they are trivially true.

Having dealt with these technical matters, we come now to the first major fact of the section.

Theorem 1.10.2. $\Omega(\mathfrak{X})$ is a one dimensional vector space over $E(\mathfrak{X})$. In particular, given any $\underline{g} \in E(\mathfrak{X}) - E_0(\mathfrak{X})$, and any $\omega \in \Omega(\mathfrak{X})$, there exists a unique $\underline{f} \in E(\mathfrak{X})$ such that $\omega = \underline{f}\, d\underline{g}$.

Proof. By (1.6.6) there exists $\underline{g} \in E(\mathfrak{X}) - E_0(\mathfrak{X})$; thus $0 \neq d\underline{g} \equiv \eta \in \Omega(\mathfrak{X})$, and $\Omega(\mathfrak{X}) \neq 0$. Let $\omega \in \Omega(\mathfrak{X})$ and $\underset{\sim}{U} \in \mathfrak{X}$. Since $\eta \neq 0$, $f_j \equiv \omega_j/\eta_j$ is well defined, for all $j \in J$. First we claim that $(f_j)_{j \in J}$ is a meromorphic function on $\mathfrak{X}$ relative to $\underset{\sim}{U}$. To prove this, assume first that $z_j \cdot z_k^{-1}$ is analytic on $z_k(V)$, V being a component of $U_j \cap U_k$. Then $\omega_k \mid V$ (resp. $\eta_k \mid V$) is $\omega_j\, T_{jk} \mid V$ (resp. $\eta_j\, T_{jk} \mid V$); hence $f_j \mid V = f_k \mid V$. If $z_j \cdot z_k^{-1}$ is anti-analytic then $\omega_k \mid V$ (resp. $\eta_k \mid V$) is $\overline{\omega_j\, T_{jk}} \mid V$ (resp. $\overline{\eta_j\, T_{jk}} \mid V$); hence $f_j \mid V = \omega_j\, T_{jk}/\eta_j\, T_{jk} \mid V = \overline{\omega}_k/\overline{\eta}_k \mid V = \overline{f}_k \mid V$, proving the theorem.

If $0 \neq \omega \in \Omega(\mathfrak{X})$ and $x \in X$, then we define the order of ω at x, $\nu_x(\omega)$, to be the order of the function $\omega_U \cdot z^{-1}$ at $z(x)$, where (U,z) is a dianalytic chart at x. It is easily checked that this is independent of the choice of (U,z).

Proposition 1.10.3. Let ω and $\eta \in \Omega(\mathfrak{X})$, $x \in X$, and let $\underline{f} \in E(\mathfrak{X})$; then $\nu_x(\omega \pm \eta) \geq \min(\nu_x(\omega), \nu_x(\eta))$ -- equality occuring if $\nu_x(\omega) \neq \nu_x(\eta)$ -- and $\nu_x(\underline{f}\omega) = \nu_x(\underline{f}) + \nu_x(\omega)$.

This may be checked locally where it is evidently true.

A meromorphic differential ω on $\mathfrak{X}$ will be called holomorphic, or of the first kind, if $\nu_x(\omega) \geq 0$ for all $x \in \mathfrak{X}$. Let $\Omega_1(\mathfrak{X})$ be the $E_0(\mathfrak{X})$- space of all such differentials. If $\nu_x(\omega) < 0$, ω is said to have a pole at x.

We now consider the question of integrating differentials along paths in X. By an arc we shall mean a continuous map from a closed interval to X, by a curve we shall mean a continuous map of the unit circle to X. We assume that all of our arcs and curves come equipped with a fixed orientation.

We first work locally. Let (U,z) be a dianalytic chart on $\mathfrak{X}$, Γ an arc or curve contained in U, and ω a differential with no poles on Γ. Then the integral

$$\int_{z(\Gamma)} \omega_U \cdot z^{-1} dz$$

is well defined [R]. If (V,w) is another dianalytic chart with $\Gamma \subset V$, then, carrying out the standard change of variable techniques, we see that the integrals are the same when $z \circ w^{-1}$ is analytic on Γ, and that

$$\int_{z(\Gamma)} \omega_U \cdot z^{-1} dz = \kappa \circ \int_{w(\Gamma)} \omega_V \cdot w^{-1} dw$$

when $z \circ w^{-1}$ is anti-analytic on Γ. Hence the integral of a meromorhphic differential is locally well defined up to complex conjugation.

Theorem 1.10.4. Let Γ be an oriented arc or curve in X, and let $\omega \in \Omega(\mathfrak{X})$ have no poles on Γ.

(a) The real part of $\int_\Gamma \omega$ is always well-defined.

(b) If Γ is an arc, or a curve, with an orientable neighborhood in X, then $\int_\Gamma \omega$ is well-defined up to complex conjugation.

(c) If $\mathfrak{X}$ is a Riemann surface then $\int_\Gamma \omega$ is a well-defined complex number.

Proof. Since the integral is locally well-defined up to complex conjugation, its real part is globally well defined. If Γ is an arc or a curve with an orientable neighborhood then we may partition it into a sequence of arcs $\Gamma_1, \ldots, \Gamma_n$ such that Γ_j is contained in U_j, where (U_j, z_j) is a dianalytic chart on X with $z_j \cdot z_{j+1}^{-1}$ analytic for $j = 1, \ldots, n-1$. This can be done in two ways, and the integral is well defined up to complex conjugation. Part (c) is, of course, classical.

Corollary 1.10.5. If $\Gamma \subset \partial X$, then $\int_\Gamma \omega$ is well defined.

Proof. The integrals are real valued.

Now let $\theta : \mathfrak{X} \longrightarrow \mathfrak{Y}$ be a non-constant morphism. We define the map $\theta^* : \Omega(\mathfrak{Y}) \longrightarrow \Omega(\mathfrak{X})$ by

$$\theta^*(\underline{f}\, d\, \underline{g}) = (\theta^*\underline{f})\, d(\theta^*\underline{g}).$$

To show that this is well defined, let $\underline{f}\, d\, \underline{g} = \underline{f}' d\underline{g}'$. Pick dianalytic charts (U,z) on $\mathfrak{X}$ and (V,w) on $\mathfrak{Y}$ such that θ maps U

homeomorphically onto V and $z = w \cdot (\theta \mid U)$. It is clear that

$$\theta^*(\underline{f}\,d\,\underline{g})_V = \theta^*(\underline{f}'d\underline{g}')_V$$

and hence that

$$\theta^*(\underline{f}\,d\,\underline{g}) = \theta^*(\underline{f}'d\underline{g}').$$

It is immediate from the definition that θ^* is an $E(\mathfrak{Y})$- linear injection of $\Omega(\mathfrak{Y})$ into $\Omega(\mathfrak{X})$. It follows that the natural map

$$E(\mathfrak{X}) \otimes_{E(\mathfrak{Y})} \Omega(\mathfrak{Y}) \longrightarrow \Omega(\mathfrak{X})$$

is an isomorphism, since both sides are one-dimensional vector spaces.

Theorem 1.10.6. Let Γ be an arc or curve in X and let $\eta \in \Omega(\mathfrak{Y})$ have no poles on $\theta(\Gamma)$. Then

$$\operatorname{Re} \int_{\theta(\Gamma)} \eta = \operatorname{Re} \int_\Gamma \theta^* \eta.$$

Proof. It clearly suffices to prove the theorem locally. So assume that we have dianalytic charts (U,z), (V,w) with $\Gamma \subset U$, $\theta(U) \subset V$. We can find an analytic function G such that the diagram

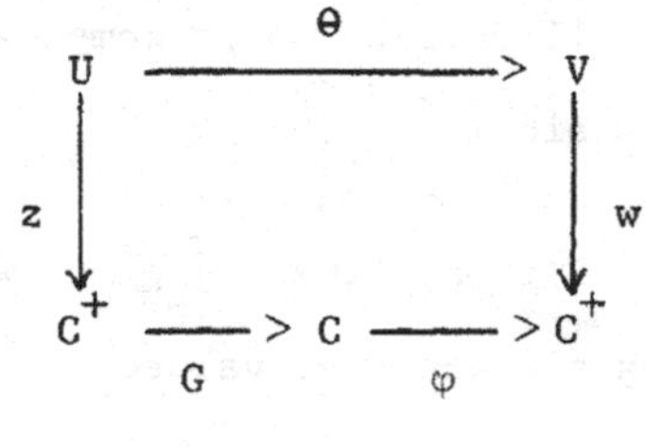

$$\begin{array}{ccccc} U & & \xrightarrow{\quad\theta\quad} & & V \\ \downarrow z & & & & \downarrow w \\ C^+ & \xrightarrow[G]{} & C & \xrightarrow[\varphi]{} & C^+ \end{array}$$

commutes. We take $\eta = h \cdot dw$, $h \in E(\mathfrak{B})$, and note that

$$(\theta^* h)_U = (\widehat{h \circ w^{-1}}) \cdot G \circ z$$

$$(\theta^* w)_U = G \cdot z,$$

where $\widehat{h \cdot w^{-1}}$ is the Schwartz extension of $h \cdot w^{-1}$ to $\varphi^{-1}(w(\mathfrak{B}))$.

Now,

$$\int_\Gamma \theta^* \eta = \int_{z(\Gamma)} (\theta^* h \cdot d\theta^* w)_U \cdot z^{-1}\, dz$$

$$= \int_{z(\Gamma)} [\widehat{(h \cdot w^{-1})} \circ G] \cdot (DG) dz$$

$$= \int_{Gz(\Gamma)} \widehat{h \cdot w^{-1}}\, d\,Gz,$$

while $\int_{\theta\Gamma} \eta = \int_{w\theta\Gamma} \widehat{h \cdot w^{-1}}\, dw$. Since $w \cdot \theta = \varphi \cdot G \cdot z$, we see that the two integrals agree locally up to complex conjugation, and hence their real parts are equal.

Example 1.10.1. Let $\mathfrak{X} = C$, $\mathfrak{Y} = C^+$, $\varphi : X \longrightarrow Y$ be the folding map. Let $z : C \longrightarrow C$, $w : C^+ \longrightarrow C^+$ be the identity maps. Take

$$\eta = \frac{1}{w} dw \in \Omega(\mathfrak{Y})$$

and consider the paths on X

$$\Gamma(t) = e^{2\pi i t} \qquad 0 \leq t \leq 1.$$

Then $\varphi^* \eta = \frac{1}{z} dz$, so

$$\int_\Gamma \varphi^* \eta = 2\pi i$$

while the path $\varphi\Gamma$ traces a half circle in each direction, so

$$\int_{\varphi\Gamma} \eta = 0.$$

Hence $\int_\Gamma \varphi^* \eta$ and $\int_{\varphi\Gamma} \eta$ need not coincide up to complex conjugation. Note that by deleting 0 from C and C^+ we may force η to be holomorphic.

Let us now see how much of our integration theory is preserved in case X is non-orientable and we consider its orienting double (§6).

Theorem 1.10.7. Let $\mathfrak{X}$ be a non-orientable Klein surface and let $f: \mathfrak{X}_o \longrightarrow \mathfrak{X}$ be the orienting double of $\mathfrak{X}$. Let Γ be an oriented arc or curve in X that is contained in an orientable neighborhood in X. Γ has two lifts in X_o, Γ_1 and Γ_2, which inherit an orientation from Γ (and which are disjoint). Let $\omega \in \Omega(\mathfrak{X})$ have no pole on Γ; then $f^*(\omega)$ has no pole on $\Gamma_1 \cup \Gamma_2$. $\int_{\Gamma_2} f^*(\omega) = \varkappa \cdot \int_{\Gamma_1} f^*(\omega)$ and $\int_{\Gamma_1} f^*(\omega)$ is equal to $\int_{\Gamma} \omega$, up to conjugation.

Proof. Let $x \in \Gamma$ and let (U,z) be a dianalytic chart at x. Recall that X_o was constructed in such a way that we can take U to be sufficiently small so that it has two disjoint lifts U_1 and U_2 in X_o and so that there exist analytic functions w_j on U_j, (U_j,w_j) being an alytic chart. We can further require that $f^*(z) \mid U_j = w_j$, $j = 1, 2$. With respect to (U_j,w_j), ν can be written as a function ω_j. By definition $f^*(\omega)_{U_j} = f^*(\omega_U)$ or $\varkappa \circ f^*(\omega_U)$ according as $j = 1, 2$. Assume that $\Gamma \subset U$, and let Γ_j be the lift of Γ in U_j. Clearly $\int_\Gamma \omega = \int_{\Gamma_1} f^*(\omega) = \varkappa \cdot \int_{\Gamma_2} f^*(\omega)$ (see e.g., [A_6 , (3.1)]). In the general case the arc or curve Γ under consideration may be covered by a finite number of such charts whose transition functions are analytic, proving the theorem.

Example 1.10.2. Let $0 < r < 1$ and let $\mathfrak{X} = \{z \in C : r < |z| < 1/r\}$. Let $\tau(\lambda) = -\overline{\lambda}^{-1}$ for all $\lambda \in \mathfrak{X}$, and note that τ has no fixed points and is an anti-analytic involution of $\mathfrak{X}$. Let $\mathfrak{Y} = \mathfrak{X}/\{1,\tau\}$ (§8); then $\mathfrak{Y}$ is a non-compact Möbius strip. Let $\alpha \in R$

and let Γ_α be the image of $[0,\pi]$ under the map $\theta \to e^{i(\theta+\alpha)}$. Note that the image of Γ_α under the quotient map is a Jordan curve Δ in Y. Let g_α be the image of the end points of Γ_α in Δ. For f meromorphic on X let $\sigma(f) \equiv \kappa \cdot f \cdot \tau$, and note that $\sigma(z) = -z^{-1}$; thus $w \equiv z - 1/z$ is invariant under σ. $dw = (1 + 1/z^2)dz$ is analytic on $\mathfrak{X}$ and invariant under σ. $\int_\Gamma dw = -4i \sin \alpha$; thus this number depends on the choice of α in a rather striking way. We see, therefore, that (1.10.4.a) can not be extended to say anything about the imaginary part of $\int_\Delta q^*(\omega)$, without many additional crippling hypotheses, where q is the quotient map.

We now define the residue $\mathrm{res}_x(\omega)$ of a differential $\omega \in \Omega(\mathfrak{X})$ at a point $x \in X$. It will be well defined up to complex conjugation. Let (U,z) be a dianalytic chart on $\mathfrak{X}$ with $x \in U$, $z(U) \subset C^+$. Then $\omega_U \cdot z^{-1}$ is an analytic function on $z(U)$, and its Schwartz extension $\widehat{\omega_U \cdot z^{-1}}$ to $\omega^{-1}(z(U))$ is analytic in a neighborhood of $z(x)$. We define

$$\mathrm{res}_x(\omega) = \mathrm{res}_{z(x)}\, \omega_U \cdot z^{-1},$$

and easily check that this is well defined up to complex conjugation. Further, if x is an interior point of X, then

$$\mathrm{res}_x(\omega) = \int_\Gamma \omega \text{ or } \kappa \int_\Gamma \omega$$

for all sufficiently small curves which go around x.

Theorem 1.10.8. Let $\omega \in \Omega(\mathfrak{X})$. Then $\int_\Gamma \omega = 0$ for every curve Γ in X, which does not pass through a pole of ω, if and only if $\omega \in dE(\mathfrak{X})$. [Note: We mean the well-defined part of the integral is zero.]

Proof. First assume $\omega = d\,\underline{f}$, and let Γ be a curve not passing through a pole of ω. Let $x_0 \in \Gamma$ and let (U,z), (U',z') be dianalytic charts at x_0 with $U = U'$, $z' = \bar{z}$. We compute $\int_\Gamma \omega$ by covering Γ with a sequence of dianalytic charts which overlay analytically starting with (U,z) and ending with (U',z') or (U,z), depending on whether orientation does or does not reverse around Γ. Then we easily see that if Γ' is the arc obtained by running through Γ, starting and ending at x_0, then

$$\int_{\Gamma'} d\,\underline{f} = \begin{cases} f_{U'}(x_0) - f_U(x_0), & \text{or} \\ f_U(x_0) - f_U(x_0) = 0. \end{cases}$$

In the second case $\int_\Gamma d\,\underline{f} = \int_{\Gamma'} d\,\underline{f} = 0$. In the first case

$$\int_\Gamma d\,\underline{f} = \operatorname{Re} \int_{\Gamma'} d\,\underline{f},$$

and this is zero because $f_{U'}(x_0) = \overline{f_U(x_0)}$ so $\int_{\Gamma'} d\,\underline{f}$ is purely imaginary.

We now prove the converse. First assume X is orientable and choose $x_0 \in X$, with $x_0 \in \partial X$ if $\partial X \neq \emptyset$. Let (U,z) be any dianalytic chart on $\mathfrak{X}$, let $y \in U$, and define

$$f_U(y) = \begin{cases} \int_\Gamma \omega & \text{if } y \text{ is not a pole of } \omega \\ \infty & \text{otherwise,} \end{cases}$$

where Γ is a path from x_0 to y, and the value of the integral is computed by covering Γ with a sequence of dianalytic charts which overlap analytically, ending with (U,z). The assumption guarantees that this is well defined. It is immediate from the definition of f_U that $D(f_U \cdot z^{-1}) = \omega_U \cdot z^{-1}$. Hence f_U is real on $U \cap \partial X$. We now see that $\underline{f} = (f_U) \in E(\mathfrak{X})$, and $d\underline{f} = \omega$.

Now assume X non-orientable and let $x_0 \in X$. The chief complication here is that we cannot assume $\underline{f}(x_0) = 0$, since we have only real constant functions, and $\underline{f}$ must be unique, up to a constant. Let (U_0, z_0) be a dianalytic chart at x_0, let (U_0', z_0') be as above, and let Γ_0 be an arc which begins and ends at x_0, with orientation reversing around Γ_0. Define

$$f_{U_0}(x_0) = -\frac{1}{2}\int_{\Gamma_0} \omega$$

where this integral is computed with a sequence of analytically overlapping dianalytic charts starting with (U_0, z_0) and ending with (U_0', z_0'). The assumption of the theorem, applied to Γ_0, implies that $f_U(x_0)$ is pure imaginary. Applied then to $\Gamma_0' - \Gamma_0$, where Γ_0' is any other such arc, it implies that

$$\int_{\Gamma_0} \omega = \int_{\Gamma_0'} \omega,$$

so that $f_{U_0}(x_0)$ is well defined. If (U,z) is any dianalytic chart on $\mathfrak{X}$, and $y \in U$, choose an arc Γ from x_0 to y which has an analytically overlapping sequence of dianalytic charts starting with (U_0, z_0) and ending with (U,z). To see that this is always possible, note that we can compose any arc Γ from x_0 to y with the above arc Γ_0 from x_0 to itself. Now define

$$f_U(y) = \int_\Gamma \omega + f_{U_0}(x_0)$$

where the integral is computed with respect to the above covering of

Γ. The assumption of the theorem guarantees that this is independent of the choice of Γ. To verify that $(f_U) = \underline{f} \in E(\mathfrak{X})$ we first note that if (U,z) and (V,w) overlap analytically at y, then trivially

$$f_U(y) = f_V(y).$$

If they overlap anti-analytically at y, note that to compute $f_V(y)$ we must first replace the arc Γ used to compute $f_U(y)$ by $\Gamma_0 + \Gamma$. Then

$$\begin{aligned} f_V(y) &= \int_{\Gamma_0+\Gamma} \omega + f_{U_0}(x_0) \\ &= \int_{\Gamma_0} \omega + \int_{\Gamma} \omega + f_{U_0}(x_0) \end{aligned}$$

where the first integral is computed with a sequence of charts starting with (U_0, z_0) and the second with a sequence starting with (U_0', z_0'). Hence

$$\begin{aligned} f_V(y) &= -2f_0(x_0) + \varkappa \cdot (f_U(y) - f_{U_0}(x_0)) + f_{U_0}(x_0) \\ &= \varkappa \cdot f_U(y) \end{aligned}$$

as required. It is immediate from the definition, that

$$D(f_U \cdot z^{-1}) = \omega_U \cdot z^{-1}$$

and since ω_U is real valued on $\partial X \cap U$, so is f_U. The theorem is proved.

§11. Automorphisms of Klein Surfaces

In this section we show how the automorphism group of a Klein surface $\mathfrak{X}$ can be obtained from that of its complex double $\mathfrak{X}_C$, and then apply this result to the disc and real projective plane. Let $\mathrm{Aut}\,\mathfrak{X}$ be the group of automorphisms of the Klein surface $\mathfrak{X}$. If X is orientable, let $\mathrm{Aut}^+\,\mathfrak{X}$ denote the subgroups of orientation preserving automorphisms of $\mathfrak{X}$: i.e., those which are automorphisms of the underlying complex structures of $\mathfrak{X}$. If $\mathfrak{X}$ is a Riemann surface, we denote its group of automorphisms by $\mathrm{Aut}^+\,\mathfrak{X}$ to emphasize that we mean analytic, rather than dianalytic, automorphisms.

Theorem 1.11.1. Let $(\mathfrak{X}_C, f, \sigma)$ be the complex double of the Klein surface $\mathfrak{X}$. Then

$$\mathrm{Aut}\,\mathfrak{X} \cong (\mathrm{Aut}^+\,\mathfrak{X}_C)^\sigma = \{\tau \in \mathrm{Aut}^+\,\mathfrak{X}_C \mid \sigma\tau\sigma = \tau\}.$$

Proof. Let $g \in \mathrm{Aut}\,\mathfrak{X}$. Then there is a unique automorphism $\tilde{g}$ of $\mathfrak{X}_C$ such that $f\tilde{g} = gf$. The mapping $g \longmapsto \tilde{g}$ is a monomorphism from $\mathrm{Aut}\,\mathfrak{X}$ to $\mathrm{Aut}^+\,\mathfrak{X}_C$. Since

$$f\sigma\tilde{g}\sigma = f\tilde{g}\sigma = gf\sigma = gf,$$

then $\sigma\tilde{g}\sigma = \tilde{g}$.

Now let $h \in \mathrm{Aut}^+\,\mathfrak{X}_C$ with $\sigma h \sigma = h$. Then h induces a continuous map h/σ from $X = X_C/\sigma$ to itself. By (1.4.3), this map is a morphism. Further $\widetilde{(h/\sigma)} = h$.

Lemma 1.11.2. Let $\mathfrak{X}$ be an orientable Klein surface. Then either $\mathrm{Aut}^+\,\mathfrak{X} = \mathrm{Aut}\,\mathfrak{X}$, or $[\mathrm{Aut}\,\mathfrak{X} : \mathrm{Aut}^+\,\mathfrak{X}] = 2$.

Proof. The product of two antianalytic maps is analytic.

We now turn to the case of compact surfaces of genus 0. If $f \in \mathrm{Aut}^+ \Sigma$, then f is a meromorphic function on Σ with a single zero and pole, so

$$f(z) = \frac{az + b}{cz + d}, \quad ad - bc \neq 0.$$

Thus $\mathrm{Aut}^+ \Sigma \cong PGL(2,C) \simeq PSL(2,C)$. Let $\varkappa$ here denote the extension of complex conjugation to Σ. Then, considering Σ as a Klein surface, $\mathrm{Aut}\, \Sigma$ is the semi-direct product of $\mathrm{Aut}^+ \Sigma$ and $\{1,\varkappa\}$:

$$\mathrm{Aut}\, \Sigma \cong \mathrm{Aut}^+ \Sigma \times \{1,\varkappa\}.$$

I.e., for $f,g \in \mathrm{Aut}^+ \Sigma$ and $\alpha,\beta \in \{1,\varkappa\}$, multiplication is given by

$$(f,\alpha)(g,\beta) = (f\alpha g \alpha, \alpha\beta).$$

Let $\mathfrak{D} = \mathrm{cl}\, C^+$ denote the disc with its canonical dianalytic structure. Since the complex double of $\mathfrak{D}$ is Σ, and the canonical antianalytic involution of Σ is $\varkappa$, we have by (1.11.1) that

$$\mathrm{Aut}\, \mathfrak{D} \cong (\mathrm{Aut}^+ \Sigma)^{\varkappa}.$$

From the isomorphism $\mathrm{Aut}^+ \Sigma \cong PGL(2,C)$ we then easily obtain

$$\mathrm{Aut}\, \mathfrak{D} \cong PGL(2,\mathbb{R}).$$

[It is not convenient to work with $PSL(2,C)$ here since an invariant element of $PGL(2,C)$ may not be represented by an invariant unimodular matrix.]

If $\begin{pmatrix} a & b \\ c & d \end{pmatrix}$ represents an element of $PGL(2,\mathbb{R})$, then the corresponding automorphism of $\mathfrak{D}$ is given by

$$z \longmapsto \varphi\left(\frac{az + b}{cz + d}\right)$$

(where φ is the folding map). This map will be analytic if and only if $(az+b)/(cz+d)$ lies in the upper half plane, which occurs when $\det\begin{pmatrix} a & b \\ c & d \end{pmatrix} > 0$. Hence $\mathrm{Aut}^+ \mathfrak{D} \cong PSL(2,\mathbb{R})$. We summarize these results below.

Theorem 1.11.3. The automorphism groups of Σ and $\mathfrak{D}$ are canonically isomorphic to the following linear groups:

$$\begin{array}{ccc} \mathrm{Aut}\,\Sigma & \cong & PGL(2,C) \times \{1,\varkappa\} \\ | & & | \\ \mathrm{Aut}^+\,\Sigma & \cong & PGL(2,C) \\ | & & | \\ \mathrm{Aut}\,\mathfrak{D} & \cong & PGL(2,\mathbb{R}) \\ | & & | \\ \mathrm{Aut}^+\,\mathfrak{D} & \cong & PSL(2,\mathbb{R}). \end{array}$$

We now turn to the real projective plane $\mathfrak{P}$. The complex double of $\mathfrak{P}$ is Σ, and the antianalytic involution is α: $\alpha(z) = -1/\bar{z}$. Thus $\mathrm{Aut}\,\mathfrak{P} \cong \mathrm{Aut}^+\,\Sigma^\alpha \cong PGL(2,C)^\alpha$. If $f \in \mathrm{Aut}^+\,\Sigma$ is represented by the matrix $\begin{pmatrix} a' & b' \\ c' & d' \end{pmatrix}$, then $\alpha f \alpha$ is represented by

$$\begin{pmatrix} -\bar{d}' & \bar{c}' \\ \bar{b}' & -\bar{a}' \end{pmatrix}$$

Hence if $\alpha f \alpha = f$, then there exists $e \in C$ with $c' = e\bar{b}'$, $d' = -e\bar{a}'$. Multiplying the matrix representing f by $(-e)^{-1/2}$, we obtain a matrix in the form

$$A = \begin{pmatrix} a & b \\ -\bar{b} & \bar{a} \end{pmatrix} .$$

Since $\det A > 0$, we may multiply by a real scalar to get $\det A = 1$. Then A is a unitary matrix, and every unimodular unitary matrix is of this form. Hence $\mathrm{Aut}\,\mathfrak{P} \cong PSU(2)$, and the isomorphism gives us the following result.

Theorem 1.11.4. The automorphism groups of Σ and $\mathfrak{P}$ are canonically isomorphic to linear groups as follows:

$$\begin{array}{ccc} \mathrm{Aut}\,\Sigma & \cong & PSL(2,C) \times \{1,\alpha\} \\ | & & | \\ \mathrm{Aut}^+\,\Sigma & \cong & PSL(2,C) \\ | & & | \\ \mathrm{Aut}\,\mathfrak{P} & \cong & PSU(2). \end{array}$$

<u>Remarks</u>. Note that when working with $\mathfrak{P}$ it is convenient to take $\mathrm{Aut}^+\,\Sigma = PSL(2,C)$, while with $\mathfrak{D}$ we took $\mathrm{Aut}^+\,\Sigma = PGL(2,C)$.

CHAPTER 2. REAL-ALGEBRAIC FUNCTION FIELDS

§1. Algebraic function fields recalled

Let k be a field. Recall that an algebraic function field (of one variable) over k is a field extension K/k such that there exists $x \in K$, transcendental over k, for which K is a finite algebraic extension over $k(x)$. (See, e.g., Chevalley [C] as a general reference on algebraic function fields.) It is easily checked that if K is an algebraic function field over k, and $y \in K$ is transcendental over k, then $[K:k(y)] < \infty$. Assume char k is 0.

K is called the field of functions and k is called a field of constants of K/k. Let $\bar{k}$ be the set of all $\alpha \in K$ such that α is algebraic over k. $\bar{k}$ is a finite extension of k and is (relatively) algebraically closed in K. $\bar{k}$ is referred to as the field of constants of K/k.

Let K/k be an algebraic function field. A subring $\mathfrak{O}$ of K is called a valuation ring of K/k if, 1) $\mathfrak{O}$ is a proper subring of K, 2) $k \subset \mathfrak{O}$, and 3) for all $\alpha \in K - \mathfrak{O}$, $\alpha^{-1} \in \mathfrak{O}$. Let $\mathfrak{O}$ be such a ring. Let U be the set of units of $\mathfrak{O}$. It is an easy exercise to check that $\mathfrak{M} = \mathfrak{O}-U$ is an ideal of $\mathfrak{O}$, and hence it is the unique maximal ideal of $\mathfrak{O}$. By definition the following sequence is k-exact: $0 \to \mathfrak{M} \to \mathfrak{O} \to \mathfrak{O}/\mathfrak{M} \to 0$. $\mathfrak{O}/\mathfrak{M} = \Omega$, an extension of k, is known as the residue class field of $\mathfrak{O}$.

Let A be an integral domain. Let $p(X) = X^n+a_1X^{n-1}+\ldots+a_n \in A[X]$; such a polynomial is called monic. A is said to be integrally closed in its quotient field F if every $\alpha \in F$ which satisfies a monic polynomial in $A[X]$, is in A. It is easy verification that every valuation ring $\mathfrak{O}$ is integrally closed in its field of fractions. Now, every $\alpha \in \bar{k}$, the field of constants, is algebraic over k, and hence satisfies a monic polynomial in $\mathfrak{O}[X]$. Since $\mathfrak{O}$ is integrally closed, $a \in \mathfrak{O}$. Thus we see that $\bar{k} \subset \mathfrak{O}$, and thus $\mathfrak{O}$ is a valuation ring of $K/\bar{k}$, not merely of K/k. Let $\mathfrak{p}$ be the canonical $\bar{k}$-homomorphism of $\mathfrak{O}$ onto Ω $(=\mathfrak{O}/\mathfrak{M})$. $\mathfrak{p}$ is known as the <u>place associated with</u> $\mathfrak{O}$. It is usual to extend $\mathfrak{p}$ to K by letting assume the "value" ∞ at all $a \in K-\mathfrak{O}$; thus $\mathfrak{p}$ extended, maps K into $\Omega \cup \{\infty\}$. Before proceeding with an example, the notion of localization should be recalled.

Let P be a (proper) prime ideal in an integral domain A. Let A be imbedded in its quotient field F. Let $A_p = \{a/b : a \in A$ and $b \in A - P\}$. Noting that $A - P$ is closed under multiplication it is then easy to check that A_p is an integral domain with a unique maximal ideal $P^e = PA_p = \{a/b : a \in P$ and $b \in A - P\}$. The situation is even more pleasant if $A = k[X]$ and P is a maximal ideal; then $P = (p)$, where p is irreducible. Further A_p is a valuation ring. Indeed, let $f \in F - A_p$. Then $f = a/b$, where $a,b \in A$, a and b have no common irreducible divisors, and p divides b. Then p necessarily does not divide a, and hence $f^{-1} = b/a \in P^e$, proving that A_p is a valuation ring.

Returning briefly to the general case, let K/k be an algebraic function field. Let $\mathrm{Riem}_k K = \{\mathfrak{O} : \mathfrak{O}$ a valuation ring of K over $k\}$. We will occasionally refer to this set as the <u>formal Riemann surface</u> of K/k.

Example 2.1.1. Let specm $k[X]$ denote the set of all maximal ideals of $k[X]$, and let λ_X map specm $k[X]$ into $\mathrm{Riem}_k k(X)$ as follows: $M \to k[X]_M$. Similarly, let $\lambda_{X^{-1}}$ map specm $k[X^{-1}]$ into $\mathrm{Riem}_k k(X)$. Now let $\mathfrak{O} \in \mathrm{Riem}_k k(X)$. Since $\mathfrak{O}$ is a valuation ring of $k(X)/k$, $X \in \mathfrak{O}$ or $X^{-1} \in \mathfrak{O}$; assume the former. Then $k[X] \subset \mathfrak{O}$ and $\mathfrak{M} \cap k[X] = P$ is a prime ideal in $k[X]$. Note that $k[X]_P \subset \mathfrak{O}$. Were P the zero ideal, then $k(X) = k[X]_P \subset \mathfrak{O}$, which is absurd. Hence P

is a non-zero proper ideal in $k[X]$, and hence a maximal ideal. But it is easily checked that $k[X]_p$, being a valuation ring with value group isomorphic to $\mathbb{Z}$, is a maximal proper subring of $k(X)$, and hence $\mathfrak{O} = k[X]_p$. This proves that λ_X is injective and maps specm $k[X]$ $\{\mathfrak{O} \in \mathrm{Riem}_k k(X) : X \in \mathfrak{O}\}$. Similarly, $\lambda_{X^{-1}}$ is an injection of specm $k[X^{-1}]$ onto $\{\mathfrak{O} \in \mathrm{Riem}_k k(X) : X^{-1} \in \mathfrak{O}\}$; thus λ_X (specm $k[X]) \cup$ $\lambda_{X^{-1}}$ (specm $k[X^{-1}]) = \mathrm{Riem}_k k(X)$. Let $\mathfrak{O} \in \mathrm{Riem}_k k(X)$ with $X \notin \mathfrak{O}$, so that $X^{-1} \in \mathfrak{M}$. Hence $\mathfrak{O} = \lambda_{X^{-1}}((X^{-1}))$. Then λ_X(Specm $k[X]$) consists of all of $\mathrm{Riem}_k k(X)$ except for the "point at infinity" $\infty = \lambda_{X^{-1}}((X^{-1}))$. It is further easy to check that if $M \in$ specm $k[X]$ then the residue class field of $\lambda_X(M)$ is canonically isomorphic to $k[X]/M$.

Again let K/k be an algebraic function field, let $\mathfrak{O} \in \mathrm{Riem}_k K$, and let U be the set of units of $\mathfrak{O}$. Recall that U is a subgroup of K^*, the multiplicative subgroup of non-zero elements of K. Consider the following exact sequence of Abelian groups: $1 \to U \to K^* \xrightarrow{V} G \to 0$. (Since U is a multiplicative group, 1 is its smallest subgroup. We wish to write G additively, and hence its smallest subgroup is 0. Hence $V(fg) = V(f) + V(g)$ for all $f,g \in K^*$.) Now let $G_+ = V(K^* \cap \mathfrak{O})$. Since $\mathfrak{O}$ is closed under multiplication, $G_+ + G_+ \subset G_+$. Let $g \in G$ and let $f \in K^*$ such that $V(f) = g$. If $f \in \mathfrak{O}$, $g \in G_+$. If $f \notin \mathfrak{O}$, then $f^{-1} \in \mathfrak{O}$, and $-g = V(f^{-1}) \in G_+$. Thus G is totally a ordered group under the order induced by G_+. Let V extend to K by taking 0 to ∞, a new element greater than all $g \in G$. V is known as the valuation of $\mathfrak{O}$ and G as its value group. It is easy to check that given any $a,b \in K$, $V(a+b) \geq \min(V(a),V(b))$, equality holding if $V(a) \neq V(b)$.

Example 2.1.2. Let $M \in$ specm $k[X]$ and let $M = (m)$; then m is irreducible in $k[X]$. Let $f \in k(X)^*$, and note that $f = a/b$, $a,b \in k[X]$, having no irreducible factors in common. $a = a_0 m^{\alpha}$ and $b = b_0 m^{\beta}$, where a_0 and b_0 are in $k[X]$, not having m as a factor, and α and β are non-negative integers. Then let $\mathfrak{O} = \lambda_X(M)$ and note that a_0 and $b_0 \in U$, and as a consequence, $V(f) = (\alpha-\beta)\cdot V(m)$, from which we see that the value group G in this case is the infinite cyclic group generated by $V(m)$. Hence the value group of each element in $\mathrm{Riem}_k k(X)$ is order isomorphic to the group $\mathbb{Z}$ of integers.

Let K/k be an algebraic function field and let L/K be a finite separable algebraic extension of degree n. Let $\iota : K \to L$ be the inclusion map. Then L/k is an algebraic function field. Let $\mathfrak{O} \in \mathrm{Riem}_k L$ and let $\iota^*(\mathfrak{O}) = \mathfrak{O} \cap K$. It is very easy to see that ι^* maps $\mathrm{Riem}_k L$ into $\mathrm{Riem}_k K$. Applying the place extension theorem (see e.g., Lang [L_1]) we have the following.

Theorem 2.1.1. $\iota^* : \mathrm{Riem}_k L \to \mathrm{Riem}_k K$ is a surjection.

Let $\mathfrak{O} \in \mathrm{Riem}_k L$, $\iota^*(\mathfrak{O}) = \mathfrak{O}'$, and let Ω (resp. Ω'), V (resp. V'), and G (resp. G') be the residue class field, valuation, and value group of $\mathfrak{O}$ (resp. $\mathfrak{O}'$) respectively. Consider the following exact sequence of k-algebras and k-homomorphisms.

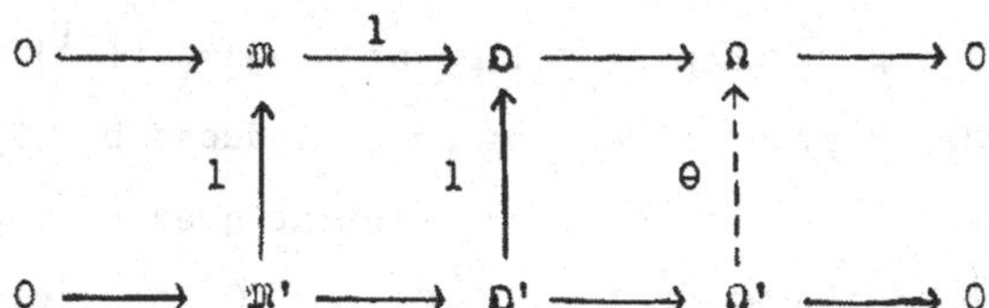

$$\begin{array}{ccccccccc} 0 & \longrightarrow & \mathfrak{M} & \xrightarrow{1} & \mathfrak{O} & \longrightarrow & \Omega & \longrightarrow & 0 \\ & & \uparrow 1 & & \uparrow 1 & & \uparrow \theta & & \\ 0 & \longrightarrow & \mathfrak{M}' & \longrightarrow & \mathfrak{O}' & \longrightarrow & \Omega' & \longrightarrow & 0 \end{array}$$

Clearly θ is a k-monomorphism of Ω' into Ω. It can easily be seen (see e.g., Zariski-Samuel [ZS_2, p. 26]), that $[\Omega:\theta(\Omega')] = f \leq n$. The integer f is known as the <u>relative degree</u> of $\mathfrak{O}$ over $\mathfrak{O}'$. Now consider the following exact sequences of Abelian groups.

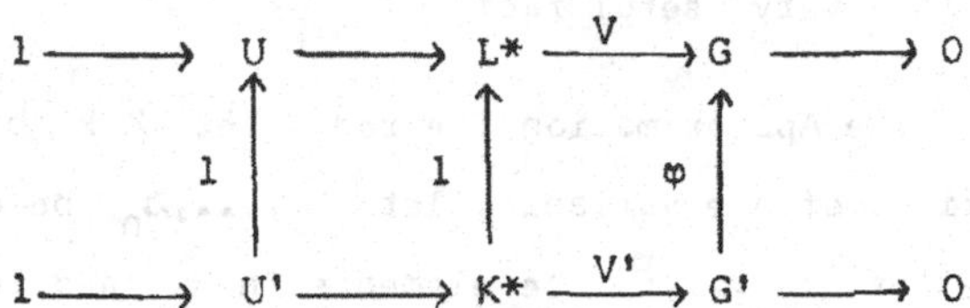

φ then is an order preserving monomorphism of G' into G. Let e be the index of $\varphi(G')$ in G. It is easily seen (see e.g., [ZS $_2$, p. 52]) that e is finite and that $e \leq n$. e is called the <u>ramification index</u> of $\mathfrak{O}$ over $\mathfrak{O}'$. Let $K = k(f) \cong k(X)$, f an element in L transcendental over k. It can now be easily shown that G is isomorphic to Z by noting that $G' \cong Z$ and, since G is a totally ordered group in which $\varphi(G')$ has finite index. Let $p \in \mathfrak{O}$ with $V(p) = 1$. Then it is easily seen that p generates the maximal ideal of $\mathfrak{O}$. The following theorem is called the "fundamental equality".

Theorem 2.1.2. Let $\mathfrak{O} \in \mathrm{Riem}_k K$. Then $\mathcal{L}*^{-1}(\mathfrak{O})$ is a finite set: $\mathcal{L}*^{-1}(\mathfrak{O}) = \{\mathfrak{O}_1,\dots,\mathfrak{O}_t\}$. Let f_j be the relative degree of $\mathfrak{O}_j$ over $\mathfrak{O}$ and e_j the ramification index of $\mathfrak{O}_j$ over $\mathfrak{O}$. Then $\Sigma_{j=1}^{t} e_j f_j = n(=[L:K])$.

(See e.g., [ZS $_2$] for a proof.)

Let $\mathbb{K}$ be the category whose objects are algebraic function fields, of one variable, over k and whose morphisms are k-monomorphisms.

Let $K \xrightarrow{\theta} L$ be a morphism in $\mathbb{K}$. Then L is a finite extension of $\theta(K)$. Given $\mathfrak{O} \in \mathrm{Riem}_k L$, let $\theta^*(\mathfrak{O}) = \theta^{-1}(\mathfrak{O} \cap K)$. Then, as noted above, θ^* maps $\mathrm{Riem}_k L$ onto $\mathrm{Riem}_k K$ and is subject to the fundamental equality (2.1.2). Let $\theta^* = \mathrm{Riem}_k \theta$; then Riem_k is a contravariant functor on ⓚ into the category of sets and maps.

The following is a very useful fact.

Theorem 2.1.3. The Approximation Theorem. Let K/k be an algebraic function field of one variable, let $\mathfrak{D}_1,\dots,\mathfrak{D}_n$ be distinct points in $\mathrm{Riem}_k K$, let $f_1,\dots,f_n$ be elements in K and let $m_1,\dots,m_n \in Z$. There exists $f \in K$ such that for all j, $1 \leq j \leq n$, $V_j(f-f_j) = m_j$, where V_j is the valuation of $\mathfrak{D}_j$ taking values in Z.

(See [ZS_2] for details.)

§2. Introduction to Complex and Real Algebraic Function Fields

Let Ⓒ be the category of algebriac function fields over C, as described in §1. The fact that C is algebraically closed leads to great simplification as will be seen below.

Example 2.2.1. Σ. specm $C[X]$ is merely $\{(X-\alpha) : \alpha \in C\}$. It is natural to identify α and $(X-\alpha)$ and thus think of specm $C[X]$ as C. Further, each residue class field of $C[X]$ is C-isomorphic to C. λ_X injects C (= specm $C[X]$) into $\Sigma \equiv \mathrm{Riem}_C\, C(X)$; let $\lambda_X(\alpha) \equiv \alpha$. Let $\lambda_{X^{-1}}(0) \equiv \infty$; then Σ is the disjoint union of C and $\{\infty\}$. Let $\beta \in C$, $\beta \neq 0$. $(X^{-1} - \beta)$, or β, is in $C[X^{-1}]$. Then $\lambda_{X^{-1}}((X^{-1} - \beta)) = \lambda_X((1 - \beta X)) = \lambda_X((\beta^{-1} - X))$; thus $\lambda_{X^{-1}}(\beta) = \lambda_X(\beta^{-1}) = \beta^{-1}$.

Let $L \in$ Ⓒ. Since C is algebraically closed, C is the field of constants in L. If $f \in L - C$, we have the following morphism in Ⓒ: $C(f) \xrightarrow{\iota} L$. Then ι^* maps $X \equiv \mathrm{Riem}_C\, L$ onto $\Sigma \equiv \mathrm{Riem}_C\, C(f)$. Let ι^*, which depends on the choice of f, be denoted by $\hat{f}$; thus each $f \in L - C$ induces a map $\hat{f}$ of X onto Σ. Let $\sigma \in \Sigma$ and let $\{x_1, \ldots, x_t\} = \hat{f}^{-1}(\sigma)$. Since C is algebraically closed, all of the residue class fields of valuation rings of L are C. Thus the fundamental equality reduces to $\sum_{j=1}^{t} e_j = n$, where e_j is the ramification index of x_j over σ and $n = [L : C(f)]$, which is also defined as the <u>degree of</u> f.

Let us now consider the category Ⓡ, of algebraic function fields over R. We begin with the following example.

Example 2.2.2. cl C^+. Given $M \in$ specm $R[X]$ it is of one of two forms: $M = (X - r)$, $r \in R$, or $M = (X^2 + bX + c)$, $b,c \in R$, and $b^2 - 4c < 0$. The former will be called <u>real points</u> and the latter <u>complex points</u> of specm $R[X]$. $r \longrightarrow (X - r)$ injects R into specm $R[X]$, and covers all real points. $\alpha \longrightarrow ((X-\alpha)(X-\bar{\alpha}))$ injects $C^+ - R$ into specm $R[X]$; thus under these maps C^+ is injected onto specm $R[X]$, with which it will be identified. λ_X injects C^+ into $\text{Riem}_R\ R(X)$; let us identify the points of C^+ with their images under λ_X. Let $\lambda_{X^{-1}}(0) \equiv \infty$; then $\text{Riem}_R\ R(X)$ is the disjoint union of C^+ and $\{\infty\}$, i.e., $\text{Riem}_R\ R(X) = \text{cl}\ C^+$ (or Δ).

Example 2.2.3. $\varphi_{\text{cl}\ C^+}$. Consider the following morphism in Ⓡ: $R(X) \xrightarrow{\iota} C(X)$. Since $C(X) = R(X)(i)$, $[C(X) : R(X)] = 2$. Let $\tau \in \text{Riem}_R\ R(X)$ $(= \text{cl}\ C^+)$ and let $\sigma \in \text{Riem}_R\ C(X)$ $(= \Sigma)$ such that $\iota^*(\sigma) = \tau$. If τ is a complex point, then no residue class field extension occurs as we pass from τ up to σ; thus the relative degree of σ over τ is 1. The minimal monic polynomial of i over R is $t^2 + 1$ and splits into $(t - i)(t + i)$ in $C[t]$. The derivative of $m(t) = t^2 + 1$, with respect to t is $2t$, and $m'(i) = 2i$. For all $\mathfrak{O} \in \text{Riem}_R\ C(X)$, $m'(i) \notin \mathfrak{M}$, $\mathfrak{M}$ being the maximal ideal of $\mathfrak{O}$; thus each point of Σ is unramified over its image in cl C^+. (See e.g., [C] for details.) Hence by (2.1.2) there exist distinct points σ_1 and σ_2 in Σ such that $\iota^*(\sigma_j) = \tau$, $j = 1, 2$. Now let τ be a real point in cl C^+. Since each residue class field of $C(X)$ is complex the relative degree of σ over τ is 2 $(= [C:R])$. Employing the fundamental equality (2.1.2) again, we see that σ is the only point over τ and that $\sigma \in R \cup \{\infty\}$. Again let τ be a complex point of cl C^+; then $\tau \in C^+$ and $\lambda_X((X^2 - (\tau + \bar{\tau})X + \tau\bar{\tau}))$ $= \tau$. Now in $C[X]$ this polynomial factors into $(X - \tau)(X - \bar{\tau})$; thus the maximal ideal $(X^2 - (\tau + \bar{\tau})X + \tau\bar{\tau})$ in $R[X]$, when extended to $C[X]$, is $(X - \tau)(X - \bar{\tau})$. Hence $\lambda_X((X - \tau)) = \tau$ and $\lambda_X((X - \bar{\tau})) = \bar{\tau}$ in $\mathrm{Riem}_R\, C(X)$ lie over τ in $\mathrm{Riem}_R\, R(X)$: i.e., $\{\sigma_1, \sigma_2\} = \{\tau, \bar{\tau}\}$ To summarize, the map ι^* of Σ onto cl C^+ is the identity map on $R \cup \{\infty\}$, and the set $(\iota^*)^{-1}(\tau) = \{\tau, \bar{\tau}\}$. We therefore see that ι^* in this context, is the folding map, $\varphi_{\text{cl } C^+}$, introduced in 1.3.

Now let E be a real-algebraic function field and let $X \equiv \mathrm{Riem}_R\, E$. Let $\bar{R}$ be the field of constants of E, which is either R, or is R-isomorphic to C. Now let $f \in E - \bar{R}$. We have seen that $R(f) \xrightarrow{\iota} E$ induces a map $\iota^* = \hat{f}$ of X onto cl C^+. Let X have the weakest topology making all $\hat{f}$ continuous.

Proposition 2.2.1. Given $E \in \circledR$, $X \equiv \mathrm{Riem}_R\, E$ is a Hausdorff space.

Proof. Let $\mathfrak{O}$ and $\mathfrak{O}'$ be distinct points in X. Since $\mathfrak{O} \neq \mathfrak{O}'$ there exists $f \in E$ which is in one of the maximal ideals of these valuation rings. Without loss of generality we may assume that $f \in \mathfrak{M}$ and $f \notin \mathfrak{M}'$. Thus $\hat{f}(\mathfrak{O}) = 0$ and $\hat{f}(\mathfrak{O}') \equiv \sigma \in \text{cl } C^+$ is not zero. Since Σ is Hausdorff there exists disjoint open sets U and V in Σ such that $0 \in U$ and $\sigma \in V$. Then $\hat{f}^{-1}(U)$ and $\hat{f}^{-1}(V)$ are disjoint open sets in X, containing $\mathfrak{O}$ and $\mathfrak{O}'$ respectively, showing that X is also Hausdorff, proving the proposition.

Let $K \xrightarrow{\iota} E$ be a morphism in $\widehat{\circledR}$.

Proposition 2.2.2. ι^* is a continuous map of X ($\equiv \mathrm{Riem}_R\, L$) onto Y ($\equiv \mathrm{Riem}_R\, K$).

Proof. As we saw in §1 of this chapter, the place extension theorem (2.1.1) shows that ι^* is surjective. Let $f \in K - \overline{R}$. Consider the following commutative diagram

(2.2.1)

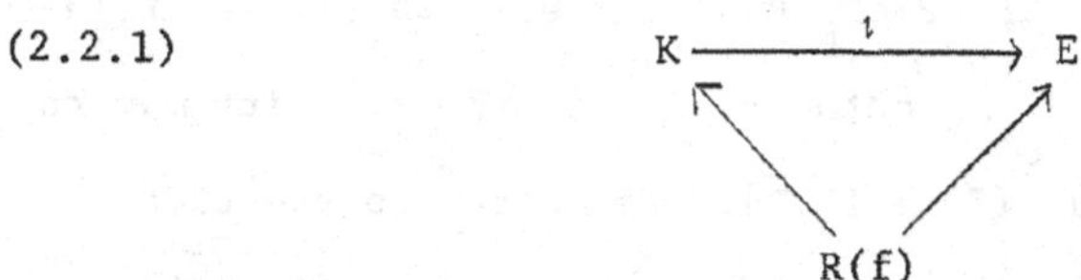

Applying Riem_R to (2.2.1) we obtain the following commutative diagram

(2.2.2)

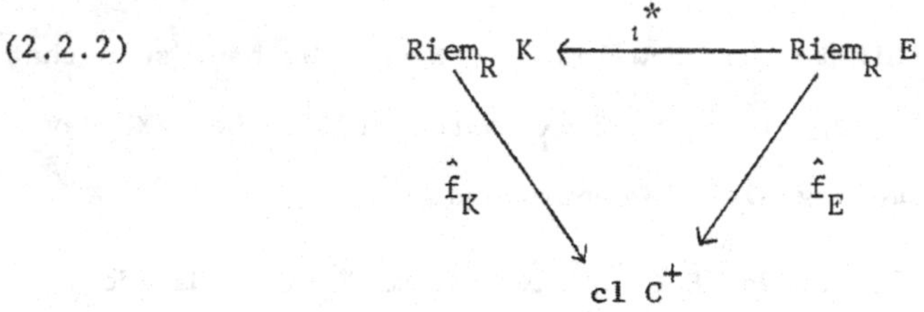

Since $\hat{f}_E$ is continuous and is $\hat{f}_K \iota^*$, ι^* is continuous, proving the proposition.

One of the main theorems about the category $\mathbb{C}$ is the following.

Theorem 2.2.3. Let $L \in \mathbb{C}$ and let $X \equiv \mathrm{Riem}_C\, L$. X is a compact surface with empty boundary. X may be given a unique complex analytic structure $\mathfrak{X}$ such that for each $f \in L$, $\hat{f}$ is a meromorphic function on $\mathfrak{X}$. $\hat{L} \equiv \{\hat{f} : f \in L\}$ is then the field of <u>all</u> meromorphic functions on $\mathfrak{X}$. Given a morphism $K \overset{\iota}{\rightarrowtail} L$ in $\mathbb{C}$, the adjoint, ι^* ($\equiv \mathrm{Riem}_C\, \iota$) of ι, is an analytic map of $\mathfrak{X}$ onto $\mathfrak{Y}$

$\equiv \mathrm{Riem}_C\, K$. That is Riem_C is a contravariant functor from Ⓒ into the category Ⓢ of all compact Riemann surfaces and non-constant analytic maps. In addition, given $x \in X$ and $f \in L - C$, which has a zero of order 1 at x, then $\hat{f}$ is a local homeomorphism at x. Further, given a compact (connected) Riemann surface $\mathfrak{X}$, $L(\mathfrak{X})$, the field of all meromorphic functions on $\mathfrak{X}$, is an object in Ⓒ; and given a morphism $\mathfrak{X} \xrightarrow{f} \mathfrak{Y}$ in Ⓢ, f^* $(\equiv L(f))$, is a C-isomorphism of $L(\mathfrak{Y})$ into $L(\mathfrak{X})$, objects in Ⓒ: i.e., L is a contravariant functor of Ⓢ into Ⓒ. Finally, up to natural equivalence, L and Riem_C are inverse functors; thus Ⓒ and Ⓢ are coequivalent categories.

See Chevelley's chapter on Riemann surfaces [C], Gunning [G], or Lang [L_1], for a proof of (2.2.3).

Let $L \in$ Ⓒ: i.e., let L be an object in Ⓒ and let $\mathfrak{O} \in \mathrm{Riem}_R\, L$. Then C is the field of constants of L, $C \subset \mathfrak{O}$; thus $\mathrm{Riem}_R\, L = \mathrm{Riem}_C\, L$, as sets. Since the folding map is continuous, the identity map of $\mathrm{Riem}_C\, L$ onto $\mathrm{Riem}_R\, L$ is a continuous bijection, thus a homeomorphism. Hence we have the following lemma.

Lemma 2.2.4. Given $L \in$ Ⓒ; then $\mathrm{Riem}_R\, L$ and $\mathrm{Riem}_C\, L$ are identical and have the same topology.

Corollary 2.2.5. Let $E \in$ Ⓡ; then $\mathrm{Riem}_R\, E$ is a compact, connected space.

Proof. Let $L \equiv E(i)$; then $L \in$ Ⓒ, and the adjoint of the injection of E into L is a continuous map of $\mathrm{Riem}_R\, L$ onto $\mathrm{Riem}_R\, E$.

Since the former is a compact, connected space, so is the latter, proving the corollary.

Thus we see that Riem_R is a contravariant functor of $\circledR$ into the category of compact, connected Hausdorff spaces and continuous surjections.

§3. Four Categories

Let $\mathfrak{K}_i$ denote the category of non-empty compact (but not necessarily connected) Klein surfaces, with and without boundary, which have a finite number of components. Let $\mathfrak{X}$ and $\mathfrak{X}'$ be objects in $\mathfrak{K}_i$, and let f be a continuous map of X onto X'. f will be called a morphism in $\mathfrak{K}_i$ if it is a morphism on each component of $\mathfrak{X}$; thus $\mathfrak{K}$ is a sub-category of $\mathfrak{K}_i$.

Let $\mathfrak{S}_i$ be the category of pairs $(\mathfrak{Y},\sigma)$, where $\mathfrak{Y}$ is a compact (but not necessarily connected) Riemann surface (with $\partial Y = \emptyset$), having a finite number of components, and σ is an antianalytic involution of $\mathfrak{Y}$. A morphism θ from $(\mathfrak{Y},\sigma)$ to $(\mathfrak{Y}',\sigma')$ is an analytic map θ of $\mathfrak{Y}$ onto $\mathfrak{Y}'$ such that $\theta\sigma = \sigma'\theta$. (The rationale for the use of the subscript i is that $\mathfrak{S}_i$ is a category of surfaces with involution.)

Now we wish to define covariant functors D and Q between these categories, $\mathfrak{K}_i \overset{D}{\rightsquigarrow} \mathfrak{S}_i \overset{Q}{\rightsquigarrow} \mathfrak{K}_i$, and show that they establish an equivalence of categories. Given $\mathfrak{X} \in \text{ob } \mathfrak{K}_i$, let $\mathfrak{X}^{(1)}$, ..., $\mathfrak{X}^{(n)}$ be its several components. Let $D(\mathfrak{X})$ be the disjoint union $\mathfrak{X}_C$ of $\mathfrak{X}_C^{(1)}, \ldots, \mathfrak{X}_C^{(n)}$, the complex doubles of $\mathfrak{X}^{(1)}, \ldots, \mathfrak{X}^{(n)}$ respectively, together with the natural antianalytic involution σ on this surface. (See Theorem 1.6.1 for details.) Note: even if $\mathfrak{X}$ is connected $\mathfrak{X}_C$ need not be (see (1.6.3) for details). Let f be the union of the natural projection of $\mathfrak{X}_C^{(1)}, \ldots, \mathfrak{X}_C^{(n)}$ onto $\mathfrak{X}^{(1)}, \ldots, \mathfrak{X}^{(n)}$ respectively (1.6.1); then $f\sigma = f$. Let $\mathfrak{X} \xrightarrow{g} \mathfrak{X}'$ be a morphism in $\mathfrak{K}_i$. By (1.6.2), there exists a unique analytic map ρ of $\mathfrak{X}_C$ into $\mathfrak{X}_C'$ which makes the following diagram commute.

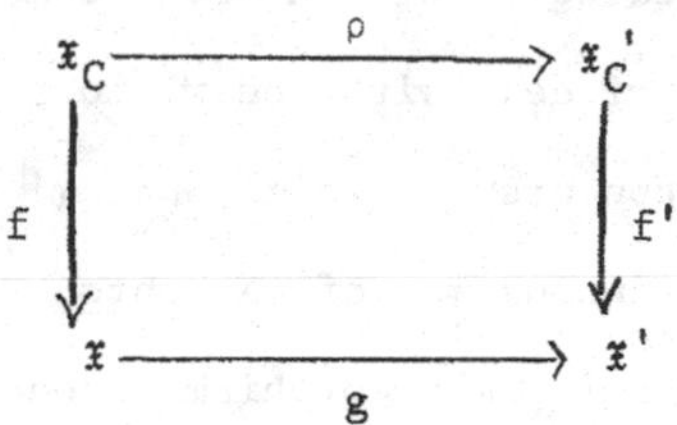

Since f and g are surjective, so is ρ. It is easily checked that $\rho\sigma = \sigma'\rho$. Let $D(g) \equiv \rho$. It is easily seen, by applying (1.6.2), that D is a covariant functor of $Ⓚ_i$ into $Ⓢ_i$. (N.b., even if X is connected, D(X) may have two components; hence we are forced to deal with disconnected surfaces here.)

Let $(\mathfrak{Y},\sigma) \xrightarrow{\rho} (\mathfrak{Y}',\sigma')$ be a morphism in $Ⓢ_i$. $\{1,\sigma\}$ is a group of dianalytic auto-homeomorphisms on Y which act discretely on Y. Let $\mathfrak{Y}/\sigma$ stand for the Klein surface $\mathfrak{Y}/\{1,\sigma\}$. (See Theorem 1.8.4 for details.) This quotient, $\mathfrak{Y}/\sigma$, will be denoted by $Q(\mathfrak{Y},\sigma)$. In (1.8.4) we also saw that there is a unique morphism π of $\mathfrak{Y}$ onto $\mathfrak{Y}/\sigma$. Since ρ maps Y onto Y' such that $\rho\sigma = \sigma'\rho$, ρ induces a set theoretic map $Q(\rho)$ of $\mathfrak{Y}/\sigma$ onto $\mathfrak{Y}'/\sigma'$. By (1.4.3), $Q(\rho)$ is a morphism.

It is an easy exercise to check that $QD(\mathfrak{X})$ is canonically isomorphic to $\mathfrak{X}$, that $DQ(\mathfrak{Y},\sigma)$ is canonically isomorphic to $(\mathfrak{Y},\sigma)$ (see (1.6.2)), and that we have indeed established an equivalence of categories.

We introduce a third category, $Ⓒ_i$, the category of pairs (A,τ), where A is a C - algebra that is C - isomorphic to a finite product of algebraic function fields in one variable over C, and τ an R-algebra involution of A such that

$$\tau(ca) = \bar{c}\tau(a), \quad \text{for all } c \in C, \text{ and all } a \in A.$$

A morphism $\gamma : (A,\tau) \longrightarrow (A',\tau')$ is a C-monomorphism $\gamma : A \longrightarrow A'$ such that $\gamma \cdot \tau = \tau' \cdot \gamma$.

We now will construct (contravariant) functors Riem_C and L between $\mathfrak{S}_i$ and $\mathfrak{C}_i$, and show that they establish (co)-equivalence of categories. Let $(\mathfrak{Y},\sigma) \in \mathfrak{S}_i$ and let $Y_1, \ldots, Y_j$ be its components. If we let $L(\mathfrak{Y})$ be the ring of meromorphic functions on $\mathfrak{Y}$, then $L(\mathfrak{Y}) \cong L(\mathfrak{Y}_1) \times \ldots \times L(\mathfrak{Y}_j)$, and that each $L(\mathfrak{Y}_j)$ is a function field in one variable over C. (See e.g. (2.2.3) for references to this cardinal classical result.) The involution $\tau \equiv L(\sigma)$, is simply defined by

$$(2.3.1) \quad \tau(f)(y) \equiv \overline{f(\sigma(y))}, \quad \text{for all } y \in Y.$$

We will let the reader complete the verification that L is a contravariant functor of $\mathfrak{S}_i$ into $\mathfrak{C}_i$.

Let $(A,\tau) \in \mathfrak{C}_i$; then $A \cong F_1 \times \ldots \times F_j$, where $F_1, \ldots, F_j$ are algebraic function fields of one variable over C. The F_j 's are uniquely determined, since they can be identified with the minimal ideals of A. Hence, for each index i, there is an index $\tau(i)$ such that $\tau(F_i) = F_{\tau(i)}$. Although an abuse of notation, define $\mathrm{Riem}_C\,(A,\tau)$ to be the pair whose first coordinate is the disjoint union of the $\mathrm{Riem}_C\, F_i$'s, $i = 1, \ldots, j$, and whose second coordinate in the involution $\sigma = \tau^*$, which takes $\mathrm{Riem}_C\, F_{\tau(i)}$ tq $\mathrm{Riem}_C\, F_i$. That L and Riem_C are inverse functors can be easily deduced from (2.2.3).

We now define a category $\textcircled{R}_i$ whose objects are R - algebras that are R - isomorphic to finite products of function fields in one variable over R, and whose morphisms are (unitary) R - algebra morphisms. We will also define functors T and F between the categories $\textcircled{C}_i$ and $\textcircled{R}_i$ which establish an equivalence of categories. Let $(A,\tau) \in \textcircled{C}_i$ and let $F(A,\tau) \equiv A^{\tau} = \{a \in A : \tau(a) = a\}$, the <u>fixed</u> R - algebra of A under the group $\{1,\tau\}$. A is R - isomorphic to $F_1 \times \ldots \times F_j$. Let $\tau(i)$ be defined such that $\tau(F_i) \equiv F_{\tau(i)}$. Now re-index, if necessary, so that

$$\tau(i) = \begin{cases} i, & i = 1, \ldots, \ell \\ i+h, & i = \ell+1, \ldots, \ell+h \\ i-h, & i = \ell+h+1, \ldots, \ell+2h . \end{cases}$$

Then it is easy to see that $F(A,\tau) \cong_R E_1 \times \ldots \times E_{\ell+h}$, where E_i is a subfield of F_i of index 2, for $i = 1, \ldots, \ell$, and $E_i \cong F_i$, for $i = \ell+1, \ldots, h$. Each E_i is then clearly a function field in one variable over R. It is easily checked that F is a covariant functor from $\textcircled{C}_i$ to $\textcircled{R}_i$.

To construct T we will use tensor products. If $B \in \textcircled{R}_i$, we define T(B) to be the pair $(C \otimes_R B, \tau)$, where $C \otimes_R B = A$ is endowed with the structure of a C - algebra in the usual way, and τ is the involution of A induced by complex conjugation on the first factor: i.e.,

(2.3.2) $\tau(c \otimes b) = \bar{c} \otimes b$.

If $B \cong_R E_1 \times \ldots \times E_j$, then $C \otimes_R B \cong_R (C \otimes_R E_1) \times \ldots$

$\times (C \otimes_R E_j)$. But $C \otimes_R E_i$ is either a function field in one variable over C (if R is relatively algebraically closed in E_i), or is R-isomorphic to $E_i \times E_i$ (if E_i contains a copy of C) [L_2]. Hence $C \otimes_R B$ is isomorphic to a finite product of function fields in one variable over C. (N.b., even if $B \cong_R E_1$, $T(B)$ may be isomorphic to $E_1 \times E_1$.)

If $B \in Ⓡ_i$, and τ is defined on $C \otimes_R B$ by (2.3.2), then it is clear that $(C \otimes_R B)^\tau$ is canonically R-isomorphic to B. If $(A,\tau) \in Ⓒ_i$, then the inclusions $C \subset A$ and $A^\tau \subset A$ induce a homeomorphism $\alpha : C \otimes_R A^\tau \longrightarrow A$. To see that α is an isomorphism, note that α is A^τ-linear, and that both $C \otimes_R A^\tau$ and A are free A^τ-modules of rank 2, with bases $\{1 \otimes 1, i \otimes 1\}$ and $\{1,i\}$, respectively. The remainder of the verification that T and F are inverse functors is a trivial exercise in multilinear algebra.

We then have constructed the following commutative diagram of categories and functors.

(2.3.3) 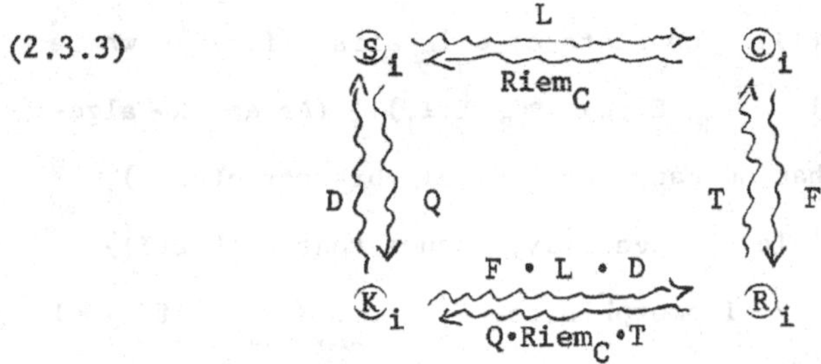

We now come to the theorem which connects the theories developed in these two chapters.

Theorem 2.3.1. The functors $F \cdot L \cdot D$ and $Q \cdot \mathrm{Riem}_C \cdot T$ establish a co-equivalence of the categories $\textcircled{K}_i$ and $\textcircled{R}_i$. The algebra $F(L(D(\mathfrak{X})))$ is a field if and only if X is connected; in this event it is R-isomorphic to $E(\mathfrak{X})$, the field of all meromorphic functions on $\mathfrak{X}$. Further, if E is an algebraic function field in one variable over R, then $Q(\mathrm{Riem}_C\,(T(E)))$ and $\mathrm{Riem}_R\,E$ are equivalent objects in $\textcircled{K}_i$.

Proof. We have just shown that D and Q establish equivalences between $\textcircled{K}_i$ and $\textcircled{S}_i$, L and Riem_C co-equivalences between $\textcircled{S}_i$ and $\textcircled{C}_i$, and F and T equivalences between $\textcircled{C}_i$ and $\textcircled{R}_i$; proving the first assertion. Assume that X is connected. As we saw in Chapter 1, §6, either $D(\mathfrak{X})$ ($\equiv \mathfrak{X}_C$, the complex double of $\mathfrak{X}$) is connected, or else it consists of two components, $\mathfrak{X}_0$ and $\mathfrak{X}_1$, each of which is mapped antianalytically onto the other by σ. Accordingly, either $E(D(\mathfrak{X}))$ is a field or it is an algebra R-isomorphic to $E(\mathfrak{X}_0) \times E(\mathfrak{X}_1)$, where $\tau(E(\mathfrak{X}_i)) = E(\mathfrak{X}_{1-i})$, $i = 0, 1$. In the first case $E(D(\mathfrak{X}))^{\tau}$ is a subfield of index two of $E(D(\mathfrak{X}))$, while in the second $E(D(\mathfrak{X}))^{\tau}$ consists of all pairs $(f, \overline{f} \cdot \sigma)$ where $f \in E(\mathfrak{X}_0)$; hence $E(D(\mathfrak{X}))^{\tau} \cong_R E(\mathfrak{X}_0) \cong_R E(\mathfrak{X}_1)$. (As an R-algebra, $(E(X_0) \times E(X_1))^{\tau}$ has no canonical C-algebra structure.) Thus $F(L(D(\mathfrak{X})))$ is a field. Conversely, assume that $F(L(D(\mathfrak{X})))$ is a field. $L(D(\mathfrak{X}))$ is R-isomorphic to $F_1 \times \ldots \times F_j$. If $j = 1$, then $D(\mathfrak{X})$ and hence $\mathfrak{X}$ is connected. Assume that $j > 1$. Assume the conventions which were in force during the initial discussion of the functor F. Then $\ell + h = 1$, and $\ell + 2h = j > 1$. As a consequence $\ell = 0$, $h = 1$, and $j = 2$, and we see that the situation described in the middle of this paragraph holds. This implies that X is connected, (orientable and without boundary).

We now will show that $F(L(D(\mathfrak{X})))$ is R-isomorphic to the field $E(\mathfrak{X})$, of all meromorphic functions on $\mathfrak{X}$. First assume that X is non-orientable or that $\partial X \neq \emptyset$. Let $(\mathfrak{X}_C, f, \sigma)$ be the complex double of X. Recall that X_C is connected. Since $f \circ \sigma = f$, $\sigma^* \circ f^* = f^*$ (1.4.9), and $f^* : E(\mathfrak{X}) \subset L(\mathfrak{X}_C)$, (1.4.9). Clearly, $f^*(E(\mathfrak{X}))$ is the fixed field of $\{1, \tau\}$: i.e., $f^*(E(\mathfrak{X})) = F(L(D(\mathfrak{X})))$. In case X is orientable and $\partial X = \emptyset$, then $\mathfrak{X}_C$ has two components, $\mathfrak{X}_0$ and $\mathfrak{X}_1$, each of which is a compact Riemann surface. $f | \mathfrak{X}_i$ is a dianalytic isomorphism; thus $E(\mathfrak{X})$ and $E(\mathfrak{X}_i)$ are R-isomorphic, $i = 0, 1$, (1.4.9). As noted above $E(\mathfrak{X}_0) \cong_R (E(\mathfrak{X}_0) \times E(\mathfrak{X}_1))^\tau$; thus $E(\mathfrak{X})$ and $F(L(D(\mathfrak{X})))$ are R-isomorphic.

Let E be an algebraic function field in one variable over R. We wish to show that $\text{Riem}_R\, E$ and $Q(\text{Riem}_C\,(T(E))$ is dianalytically equivalent. From $\iota : E \longrightarrow T(E)$ $(\equiv C \otimes_R E)$, we get a dianalytic map $\iota^* : \text{Riem}_C\, T(E) \longrightarrow \text{Riem}_R\, E$. Since $\sigma \circ \iota = \iota$, this induces a (set-theoretic) map

(2.3.4) $\iota^*/\sigma : Q(\text{Riem}_C\,(T(E))) \longrightarrow \text{Riem}_R\, E,$

which is, by (1.5.4), a morphism. By the place extension theorem, in the case T(E) is a field, ι^* is surjective. If T(E) is not a field, $T(E) \cong_R E \times E$, and by definition, ι^* is surjective. Thus ι^*/σ is surjective. Since ι^* is a double cover, and since Q is the quotient map under the involution σ^*, ι^*/σ must be an injection; thus ι^*/σ is an invertible morphism, proving (2.3.1).

Let $F \circ L \circ D$ restricted to Ⓚ, the category of connected compact Klein surfaces, possibly with boundary, be denoted by E. Let $Q \circ \mathrm{Riem}_C \circ T$ restricted to Ⓡ, the category of real algebraic function fields in one variable, be denoted by Riem_R. Theorem 2.3.1 allows us to indulge in this abuse of notation and gives us the following.

Corollary 2.3.2. E and Riem_R establish a co-equivalence of the categories Ⓚ and Ⓡ.

Let $\mathfrak{X} \xrightarrow{f} \mathfrak{Y}$ be a morphism in Ⓚ. Since $E(\mathfrak{Y}) \xrightarrow{f^* \equiv E(f)} E(\mathfrak{X})$ is a morphism in Ⓡ, the field extension $E(\mathfrak{X})/f^*(E(\mathfrak{Y}))$ has finite degree, n. Let n be known as the <u>degree of</u> f. Let $y \in Y$ and let $f^{-1}(y) = \{x_1, \ldots, x_m\}$. Let e_i be the ramification index of f at x_i (see the first paragraph of 1, §5 for a definition). If $x_i \in \partial X$ or if $y \notin \partial Y$, let $f_i \equiv 1$. However, if $y \in \partial Y$ and $x_i \notin \partial X$, let $f_i \equiv 2$. f_i will be known as the <u>relative degree</u> of x_i over y.

Theorem 2.3.3 (Fundamental equality). $\sum_{i=1}^{m} e_i f_i = n.$

Proof. By (2.3.2), $\mathfrak{X} \xrightarrow{f} \mathfrak{Y}$ may be regarded as being identical to $\mathrm{Riem}_R\, E(\mathfrak{X}) \xrightarrow{f^{**}} \mathrm{Riem}_R\, E(\mathfrak{Y})$. Note that f_i is the degree of the residue class field of y. Recalling the definition of e_i (in 1, §5), Lemma 1.5.3, and the notion of a normal form, we may choose dianalytic charts (U,z) and (V,w) at x_i and y respectively such that $f\,|\,U = w^{-1}\varphi(\pm z)^{e_i}$. Assume first that $\partial Y = \emptyset$ and that

Y is orientable. Then using diagram (1.4.1a) rather than (1.4.1), φ can be eliminated and thus $f \mid U = w^{-1}(\pm z)^{e_i}$. There exists analytic structure $\mathfrak{X}_1 \subset \mathfrak{X}$ and $\mathfrak{Y}_1 \subset \mathfrak{Y}$ (1.2.4). Thus the problem reduces to the classical case and we see that e_i is the ramification index of the valuation ring x_1 over the valuation ring y. Assume now that $\partial Y \neq \emptyset$ or that Y is non-orientable. Consider the following commutative diagram in Ⓚ,

(2.3.4)

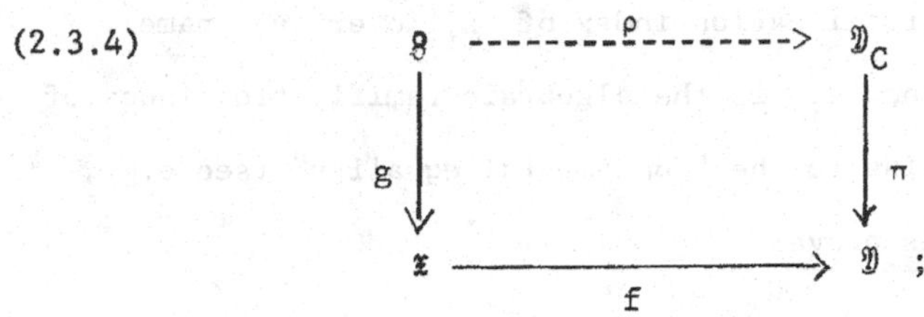

where $\mathfrak{g} = \mathfrak{X}$ and $g = \iota$ if $\partial X = \emptyset$ and X is orientable, or if not $\mathfrak{g} = \mathfrak{X}_C$ and g is the natural projection of $\mathfrak{X}_C$ onto $\mathfrak{X}$; and ρ is the unique analytic map making (2.3.4) commutative (see (1.6.2) for details.) Applying E to (2.3.4) we obtain the following commutative diagram in Ⓡ:

(2.3.5)

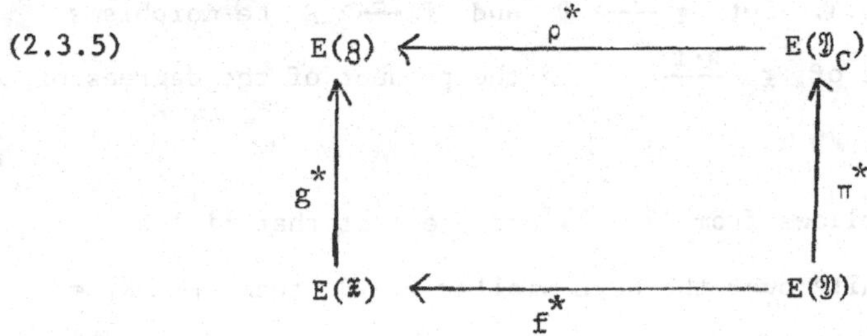

By Theorem (2.3.1), $E(\mathfrak{Y}_C) = C \otimes_R E(\mathfrak{Y})$, which in this case is $E(\mathfrak{Y})(i)$. Similarly $E(\mathfrak{g}) = E(\mathfrak{X})(i)$. Thus the vertical extensions in diagram (2.3.5) are unramified, (see e.g. [ZS_1 , p. 303]). Let

$y' \in (\pi^*)^{-1}(y)$ and let $z_i \in (\rho^*)^{-1}(y')$; then $g^*(z_i) = x_i$. As remarked the analytic ramification index e_i' of ρ at z_i is the same as the algebraic ramification index of z_i over y'. Since the vertical extensions in (2.3.5) are unramified, e_i' is the algebraic ramification index of x_i over y (see, e.g., [C , p. 53].) Returning to (2.3.4), it is clear from the construction of the complex double that the analytic ramification index of z_i over y' is the same as the dianalytic ramification index of x_i over y; namely e_i: hence $e_i' = e_i$, and e_i is the algebraic ramification index of x_i over y. On appealing to the "fundamental equality" (see e.g., (2.1.2)), the theorem is proved.

Remark. Theorem 2.3.3 is a global result about morphisms in Ⓚ which says roughly that they assume each value the same number of times. Generalizing this to non-compact Klein surfaces takes us out of a finitistic theory into Nevanlinna theory, which is beyond the scope of this monograph.

Proposition 2.3.4. Let $\mathfrak{X} \xrightarrow{f} \mathfrak{Y}$ and $\mathfrak{Y} \xrightarrow{g} \mathfrak{Z}$ be morphisms in Ⓚ; the degree of $\mathfrak{X} \xrightarrow{g \cdot f} \mathfrak{Z}$ is the product of the degrees of g and of f.

Proof. This follows from (2.3.2) and the fact that fields $K \subset L \subset M$, each finite over the next smaller field, then $[M : K] = [M : L][L : K]$.

§4. Double Coverings

Let $f : \mathfrak{X} \longrightarrow \mathfrak{Y}$ be a double cover of compact Klein surfaces. Using the results of the previous section and some elementary field theory we obtain

Lemma 2.4.1. Let $\mathfrak{X} \xrightarrow{f} \mathfrak{Y}$ be a morphism in Ⓚ, which is a double cover. There exists $\underline{r} \in E(\mathfrak{Y})$ such that $E(\mathfrak{X}) = E(\mathfrak{Y})(\sqrt{\underline{r}})$. Further, $\underline{r}$ is unique up to a non-zero square multiple.

The behavior of f is closely linked to the behavior of $\underline{r}$, as the next result chronicles.

Theorem 2.4.2 a). f is ramified at $x \in \mathfrak{X}$ if and only if $\underline{r}$ has an odd order zero or pole at $f(x) \in \mathfrak{Y}$.

b). $x \in \partial X$ if and only if $0 \leq \underline{r}(f(x)) \leq \infty$ and $f(x) \in \partial Y$.

Note: a) if $\underline{r}$ has an odd order zero or pole at $f(x)$ then so does $\underline{t}^2 \underline{r}$, when $\underline{t} \in E(\mathfrak{Y})$ and $\underline{t} \neq 0$. Further note that $0 \leq \underline{r}(f(x)) \leq \infty$ implies $0 \leq \underline{t}^2 \underline{r}(f(x)) \leq \infty$. Hence, (2.4.2) is not immediately ruled out.

Proof. a) Let $m(x) \equiv x^2 - \underline{r} \in E(\mathfrak{Y})[x]$, and note that this is the minimal monic polynomial for $\sqrt{\underline{r}}$ not in this ring. $m'(x) = 2x$, and $m'(\sqrt{\underline{r}}) = 2\sqrt{\underline{r}}$. Let $X_o \equiv X - \{x \in X : \underline{r}(f(x)) = \infty\}$; if f is ramified at $x \in X_o$ then $2\sqrt{\underline{r}}(x) = 0$: i.e., $\underline{r}(f(x)) = 0$. (See [ZS_1, Lemma 1, p. 299] for details.) If $\sqrt{\underline{r}}$ has a pole at x we can find $\underline{t} \in E(\mathfrak{Y})$ (by e.g., the approximation theorem), such that $\underline{t} \neq 0$ and $\underline{t}^2 \underline{r}$ does not have a pole at $f(x)$; thus we can assume that $x \in X_o$.

Let ν_x be the valuation at x and ν_y be the valuation at $y \equiv f(x)$, and let them have as their value group Z. Let e be the ramification index of x over y.

$$e\,\nu_y(\underline{r}) = \nu_x(\underline{r}) = \nu_x((\sqrt{\underline{r}})^2) = 2\,\nu_x(\sqrt{\underline{r}})$$

Since 2 divides $e\,\nu_y(\underline{r})$ it must either divide $\nu_y(\underline{r})$ or e. Hence odd order zeros of $\underline{r}$ correspond to ramified points. Conversely assume we are dealing with x such that $\underline{r}$ has an even order zero at f(x); then we can find $\underline{t} \in E(\mathfrak{Y})$ such that $\underline{t}^2\underline{r}$ has neither zero nor pole at f(x), and hence f is not ramified at x, proving a).

b) Possibly with the aid of $\underline{t}^2$ we may assume that $\underline{r}$ is finite at f(x). Let Ω_x and Ω_y be the residue class fields of x and y ($\equiv f(x)$) respectively. Assume that $x \in \partial X$; then $y \in \partial Y$, and $\Omega_x = R = \Omega_y$. Hence $\underline{r}(f(x)) \in R$, and since $\sqrt{\underline{r}(f(x))}$ must be in Ω_x, $0 \leq r(f(x)) \leq \infty$. Conversely assume that $0 \leq \underline{r}(f(x)) \leq \infty$ and $y \in \partial Y$. Then $\Omega_y = R$, $R(\sqrt{\underline{r}(f(x))}) = \Omega_x$, which is thus real, proving the theorem.

(2.4.2) taken together with much of the rest we know can be used in two different ways: first, given a double covering $\mathfrak{X} \xrightarrow{f} \mathfrak{Y}$, an element $\underline{r} \in E(\mathfrak{Y})$, that is not a square, a corresponding double cover $\mathfrak{X} \xrightarrow{f} \mathfrak{Y}$ exists. Let us apply the first method to the unramified doubles discussed in 1, §6, to get some feeling for the process. Note: given $\mathfrak{Y}$ we will have to restrict our attention to connected $\mathfrak{X}$, since we are now working with the category Ⓚ and not $Ⓚ_i$.

The Complex Double

Let $\mathfrak{Y}$ be non-orientable or let $\partial Y \neq \emptyset$; then the complex double $\mathfrak{Y}_C \xrightarrow{f} \mathfrak{Y}$ exists in Ⓚ. This case has been mentioned before, is obvious, but is of importance, since $E(\mathfrak{Y}_C) \cong C \hat{\otimes}_R E(\mathfrak{Y}) \cong E(\mathfrak{Y})(i)$. Hence we can choose $\underline{r} = -1$.

The Orienting Double

Let $\mathfrak{Y}$ be non-orientable and let $\mathfrak{X} = \mathfrak{Y}_o$, the orienting double of $\mathfrak{Y}$. Then $\mathfrak{X} \xrightarrow{f} \mathfrak{Y}$ is in Ⓚ, $\mathfrak{X}$ is orientable and $f^{-1}(\partial Y) = \partial X$. Then there exists $\underline{r} \in E(\mathfrak{Y})$ such that $E(\mathfrak{X}) = E(\mathfrak{Y})(\sqrt{\underline{r}})$. By (2.4.1), since f is unramified, $\underline{r}$ has no odd order zeroes or poles. Since $f^{-1}(\partial Y) = \partial X$, for all $y \in \partial Y$, $0 \leq \underline{r}(y) \leq \infty$; thus we have the following:

Corollary 2.4.3. Let $\mathfrak{Y}$ be a non-orientable compact Klein surface. There exists an element $\underline{r} \in E(\mathfrak{Y})$ (by no means unique) such that $\underline{r}$ is non-constant and not a square, $\underline{r}$ has no odd order zeros or poles, and such that for all $y \in \partial Y$, $0 \leq \underline{r}(y) \leq \infty$.

The Schottky Double

In order to apply the Schottky double to a new case, not comprehended by the complex double, assume that $\mathfrak{Y}$ is non-orientable and that $\partial Y \neq \emptyset$. Let $\mathfrak{X} = \mathfrak{Y}_S$ and consider a $\underline{q}$ corresponding to $\mathfrak{X} \xrightarrow{f} \mathfrak{Y}$. We then have

Corollary 2.4.4. Let $\mathfrak{Y}$ be a non-orientable compact Klein surface with a non-void boundary. There exists an element $\underline{q} \in E(\mathfrak{Y})$ (by no means unique) such that $\underline{q}$ has no odd order zeroes or poles, and such that for all $y \in \partial Y$, $-\infty \leq \underline{q}(y) \leq 0$.

We come then to the main theorem of this section, which generalizes a classic theorem due to Witt [W_1] in several ways.

Theorem 2.4.5. Let $\mathfrak{X}$ be a compact Klein surface, and let $I_1, \ldots, I_n$ be disjoint closed intervals or complete components of ∂X. There exists $\underline{r} \in E(\mathfrak{X})$ such that $x \in \bigcup_{j=1}^{n} I_j$ implies $0 \leq \underline{r}(x) \leq \infty$, $x' \in \partial X - \bigcup_{j=1}^{n} I_j$ implies $-\infty < \underline{r}(x) < 0$, and $\underline{r}$ has zeroes and poles of odd order at and only at the endpoints of the I_j's.

Proof. Let W' be an unramified topological double cover of X. Let the points w and w' above $x \in \overline{\partial X - \bigcup_{j=1}^{n} I_j}$ be identified, for all such x and let the resulting space be W. Using the results obtained in 1, §5, W can be given dianalytic structure $\mathfrak{W}$ such that the quotient map $\mathfrak{W} \xrightarrow{q} \mathfrak{X}$ is a morphism in $\mathbb{K}$. Since the degree of q is 2, there exists $\underline{r} \in E(\mathfrak{X})$ such that $E(\mathfrak{W}) = E(\mathfrak{X})(\sqrt{\underline{r}})$. Using (2.4.2) the result is proved.

Finally, we may now interpret (1.6.10) in terms of functions.

Theorem 2.4.6. Let $\mathfrak{X}$ be a compact Klein surface such that ∂X has $r \geq 1$ components. Let G be the subgroup of $E(\mathfrak{X})^*$ consisting of functions having no zeroes or poles of odd order. Then

$$[G : E(\mathfrak{X})^2] = 2^a$$

where $a = r + 1 - \chi(X)$.

§5. Some Elementary Applications

If $f : \mathfrak{X} \longrightarrow \mathfrak{Y}$ is a non-constant morphism of Klein surfaces, let $G = G(f)$ be the group of all automorphisms g of $\mathfrak{X}$ such that $f g = f$. Such an automorphism is known as a deck transformation of f.

Lemma 2.5.1. Let $f : \mathfrak{X} \longrightarrow \mathfrak{Y}$ be an n-fold covering. Then

(i) $|G(f)| \leq n$;

(ii) If $|G| = n$, G acts transitively on the fibers of f.

Proof. Let y be an interior point of Y which does not ramify. Then there is a neighborhood V of y with $f^{-1}(V)$ consisting of n disjoint copies of V. If $f(x) = y$, then any two elements of G which agree at x must agree in a neighborhood of x. And hence must be equal. Other points of Y are limits of sequences of unramified interior points. The lemma follows.

Proposition 2.5.2. Let $f : \mathfrak{X} \longrightarrow \mathfrak{Y}$ be a morphism in Ⓚ.

(i) $G(f)$ is anti-isomorphic to the group of all $f^*(E(\mathfrak{Y}))$ automorphisms of $E(\mathfrak{X})$.

(ii) There exists a morphism $g : \mathfrak{Z} \longrightarrow \mathfrak{X}$ in Ⓚ such that $f \cdot g$ is normal.

Proof. (i) follows from (2.3.2). (ii) is then a consequence of elementary field theory, using (2.3.2) again.

From (2.3.2), we also obtain the following result.

Proposition 2.5.3. Let $\mathfrak{X}$ be a compact Klein surface. Then Aut $\mathfrak{X}$ is anti-isomorphic to the group of all $\mathbb{R}$ - automorphisms of $E(\mathfrak{X})$.

If $\mathfrak{X}$ is a compact Klein surface, then the holomorphic differentials on $\mathfrak{X}$ form a finite dimensional vector space $\Omega(\mathfrak{X})$ over $E_o(\mathfrak{X})$, and $\dim_{E_o(\mathfrak{X})}\Omega(\mathfrak{X}) = p(\mathfrak{X})$ is known as the <u>algebraic genus</u> of $\mathfrak{X}$ (see [C]). When $E_o(\mathfrak{X}) \cong C$, then it is classical that $p(\mathfrak{X})$ equals the topological genus $1/2(2 - \chi(X))$.

Proposition 2.5.4. If $\mathfrak{X}$ is a compact Klein surface with $E_o(\mathfrak{X}) = \mathbb{R}$, then

$$p(\mathfrak{X}) = 1 - \chi(X).$$

Proof. Since the algebraic genus is invariant under separable constant field extension, $p(\mathfrak{X}) = p(\mathfrak{X}_C)$. Further $\chi(X_C) = 2\chi(X)$. Thus $p(\mathfrak{X}) = p(\mathfrak{X}_C) = 1/2(2 - \chi(X_C)) = 1 - \chi(X)$.

We can now rephrase a theorem of Witt in the language of Klein surfaces, and then deduce an interesting existence theorem for morphisms. The original theorm of Witt is the following: if X is a real algebraic curve, and f is an algebraic function which is non-negative on the real locus of X, then f is a sum of two squares of algebraic functions. In the language of Klein surfaces this becomes:

Theorem 2.5.5. If $\mathfrak{X}$ is a compact Klein surface and $f \in E(\mathfrak{X})$ is non-negative on ∂X, then f is a sum of two squares in $E(\mathfrak{X})$.

Corollary 2.5.6. If $\partial X = \emptyset$, then -1 is a sum of two squares in $E(\mathfrak{X})$.

Corollary 2.5.7. If $\partial X = \emptyset$, then there is a non-constant morphism from $\mathfrak{X}$ to the real projective plane $\mathfrak{P}$.

Proof. If $E_o(\mathfrak{X}) \cong C$, then a morphism is obtained from

$$\mathfrak{X} \longrightarrow \Sigma \longrightarrow \mathfrak{P}.$$

Otherwise, take $f,g \in E(\mathfrak{X})$ with $f^2 + g^2 = -1$. Then f,g generate a subfield of $E(\mathfrak{X})$ isomorphic to $E(\mathfrak{P})$. Applying (2.3.2) to this inclusion we obtain the corollary.

BIBLIOGRAPHY[1]

[A_1] L. Ahlfors, Open Riemann surfaces and extremal problems on compact subregions, Comment. Math. Helv. 24 (1950), 100-134.

[A_2] N. L. Alling, A proof of the corona conjecture for finite open Riemann surfaces, Bull. Amer. Math. Soc. 70 (1964), 110-112.

[A_3] _____, Extensions of meromorphic function rings over non-compact Riemann surfaces I, Math. Z. 89 (1965), 273-299.

[A_4] _____, Extensions of meromorphic function rings over non-compact Riemann surfaces II, Math. Z. 93 (1966), 345-394.

[A_5]* _____, Real Banach algebras and non-orientable Klein surfaces I, J. Reine Angew. Math. 241 (1970), 200-208.

[A_6]* _____, Analytic and harmonic obstruction on non-orientable Klein surfaces, Pacific J. Math. 36 (1971), 1-19.

[A_7]* _____, Algebraic and analytic geometry on real algebraic curves, (in preparation).

[AC]* N. L. Alling and L. A. Campbell, Real Banach algebras II, (in preparation).

[AG]* N. L. Alling and N. Greenleaf, Klein surfaces and real algebraic function fields, Bull. Amer. Math. Soc. 75 (1969), 869-872.

[AS] L. Ahlfors and L. Sario, Riemann Surfaces. Princeton University Press, Princeton, New Jersey, 1960.

[B]* L. Berzolari, Allg meine Theorie der höheren ebenen algebraischen Kurven, Encyklopädie der Math. Wiss., III, 2.1.C4.

---

[1] References marked with an asterisk are those which represent all material on Klein surfaces in the literature, which we have found.

[C] C. Chevalley, Introduction to the theory of algebraic functions of one variable, Math. Surveys, no. 6, Amer. Math. Soc., Providence, Rhode Island, 1951; Russian translation, Fizmetgiz, Moscow, 1959.

$[G_1]^*$ N. Greenleaf, Analytic sheaves on Klein surfaces, Pacific J. Math. (to appear).

$[G_2]$ R. C. Gunning, Lectures on Riemann surfaces, Princeton Mathematical Notes, Princeton University Press, 1966.

[GR]* N. Greenleaf and W. Read, Positive holomorphic differentials on Klein surfaces, Pacific J. Math. 32 (1970), 711-713.

$[G_2R]$ R. C. Gunning and H. Rossi, Analytic functions of several complex variables, Prentice-Hall, Englewood Cliffs, New Jersey, 1965.

[K]* F. Klein, Über Riemanns Theorie der algebraischen Funktionen und ihrer Integrale, Teubner, Leipzig, 1882.

$[L_1]$ S. Lang, Introduction to algebraic geometry, Interscience, New York, 1958.

$[L_2]$ ____, Algebra, Addison-Wesley, Reading, Massachusetts, 1965.

[M] W. S. Massey, Algebraic topology: an introduction, Harcourt, Brace and World, 1967.

[R] R. Redheffer, Homotopy theory of function theory, American Math. Monthly, 76 (1969), 778-787.

[SS]* M. Schiffer and D. Spencer, Functionals of Finite Riemann surfaces, Princeton University Press, Princeton, New Jersey, 1954.

[SZ] S. Saks and A. Zygmund, Analytic Functions, Monografie Matematyczne, Warsaw, 1952.

$[W_1]$ E. Witt, Zerlegung reeller algebraischer Funktionen in Quadrate Schiefkörper über reelem Funktionen köerper, J. Reine. Angew. Math. 171 (1934), 4-11.

[W_2] H. Weyl, Die Idee der Riemannschen Fläche, Leipzig, 1913.

[ZS_1] O. Zariski and P. Samuel, Commutative algebra I, van Nostrand, Princeton, New Jersey, 1958.

[ZS_2] _____, Commutative algebra II, van Nostrand, Princeton, New Jersey, 1960.

INDEX OF UNDERLINED NOTIONS

Section

Lecture Notes in Mathematics

Bisher erschienen/Already published

Vol. 38: R. Berger, R. Kiehl, E. Kunz und H.-J. Nastold, Differentialrechnung in der analytischen Geometrie IV, 134 Seiten. 1967 DM 12,–

Vol. 39: Séminaire de Probabilités I. II, 189 pages. 1967. DM 14,–

Vol. 40: J. Tits, Tabellen zu den einfachen Lie Gruppen und ihren Darstellungen. VI, 53 Seiten. 1967. DM 6,80

Vol. 41: A. Grothendieck, Local Cohomology. VI, 106 pages. 1967. DM 10,–

Vol. 42: J. F. Berglund and K. H. Hofmann, Compact Semitopological Semigroups and Weakly Almost Periodic Functions. VI, 160 pages. 1967. DM 12,–

Vol. 43: D. G. Quillen, Homotopical Algebra. VI, 157 pages. 1967. DM 14,–

Vol. 44: K. Urbanik, Lectures on Prediction Theory. IV, 50 pages. 1967. DM 5,80

Vol. 45: A. Wilansky, Topics in Functional Analysis. VI, 102 pages. 1967. DM 9,60

Vol. 46: P. E. Conner, Seminar on Periodic Maps. IV, 116 pages. 1967. DM 10,60

Vol. 47: Reports of the Midwest Category Seminar I. IV, 181 pages. 1967. DM 14,80

Vol. 48: G. de Rham, S. Maumary et M. A. Kervaire, Torsion et Type Simple d'Homotopie. IV, 101 pages. 1967. DM 9,60

Vol. 49: C. Faith, Lectures on Injective Modules and Quotient Rings. XVI, 140 pages. 1967. DM 12,80

Vol. 50: L. Zalcman, Analytic Capacity and Rational Approximation. VI, 155 pages. 1968. DM 13,20

Vol. 51: Séminaire de Probabilités II. IV, 199 pages. 1968. DM 14,–

Vol. 52: D. J. Simms, Lie Groups and Quantum Mechanics. IV, 90 pages. 1968. DM 8,–

Vol. 53: J. Cerf, Sur les difféomorphismes de la sphère de dimension trois (Γ_4 = O). XII, 133 pages. 1968. DM 12,–

Vol. 54: G. Shimura, Automorphic Functions and Number Theory. VI, 69 pages. 1968. DM 8,–

Vol. 55: D. Gromoll, W. Klingenberg und W. Meyer, Riemannsche Geometrie im Großen. VI, 287 Seiten. 1968. DM 20,–

Vol. 56: K. Floret und J. Wloka, Einführung in die Theorie der lokalkonvexen Räume. VIII, 194 Seiten. 1968. DM 16,–

Vol. 57: F. Hirzebruch und K. H. Mayer, O (n)-Mannigfaltigkeiten, exotische Sphären und Singularitäten. IV, 132 Seiten. 1968. DM 10,80

Vol. 58: Kuramochi Boundaries of Riemann Surfaces. IV, 102 pages. 1968. DM 9,60

Vol. 59: K. Jänich, Differenzierbare G-Mannigfaltigkeiten. VI, 89 Seiten. 1968. DM 8,–

Vol. 60: Seminar on Differential Equations and Dynamical Systems. Edited by G. S. Jones. VI, 106 pages. 1968. DM 9,60

Vol. 61: Reports of the Midwest Category Seminar II. IV, 91 pages. 1968. DM 9,60

Vol. 62: Harish-Chandra, Automorphic Forms on Semisimple Lie Groups X, 138 pages. 1968. DM 14,–

Vol. 63: F. Albrecht, Topics in Control Theory. IV, 65 pages. 1968. DM 6,80

Vol. 64: H. Berens, Interpolationsmethoden zur Behandlung von Approximationsprozessen auf Banachräumen. VI, 90 Seiten. 1968. DM 8,–

Vol. 65: D. Kölzow, Differentiation von Maßen. XII, 102 Seiten. 1968. DM 8,–

Vol. 66: D. Ferus, Totale Absolutkrümmung in Differentialgeometrie und -topologie. VI, 85 Seiten. 1968. DM 8,–

Vol. 67: F. Kamber and P. Tondeur, Flat Manifolds. IV, 53 pages. 1968. DM 5,80

Vol. 68: N. Boboc et P. Mustată, Espaces harmoniques associés aux opérateurs différentiels linéaires du second ordre de type elliptique. VI, 95 pages. 1968. DM 8,60

Vol. 69: Seminar über Potentialtheorie. Herausgegeben von H. Bauer. VI, 180 Seiten. 1968. DM 14,80

Vol. 70: Proceedings of the Summer School in Logic. Edited by M. H. Löb. IV, 331 pages. 1968. DM 20,–

Vol. 71: Séminaire Pierre Lelong (Analyse), Année 1967 – 1968. VI, 190 pages. 1968. DM 14,

Vol. 72: The Syntax and Semantics of Infinitary Languages. Edited by J. Barwise. IV, 268 pages. 1968. DM 18,–

Vol. 73: P. E. Conner, Lectures on the Action of a Finite Group. IV, 123 pages. 1968. DM 10,–

Vol. 74: A. Fröhlich, Formal Groups. IV, 140 pages. 1968. DM 12,–

Vol. 75: G. Lumer, Algèbres de fonctions et espaces de Hardy. VI, 80 pages. 1968. DM 8,–

Vol. 76: R. G. Swan, Algebraic K-Theory. IV, 262 pages. 1968. DM 18,–

Vol. 77: P.-A. Meyer, Processus de Markov: la frontière de Martin. IV, 123 pages. 1968. DM 10,–

Vol. 78: H. Herrlich, Topologische Reflexionen und Coreflexionen. XVI, 166 Seiten. 1968. DM 12,–

Vol. 79: A. Grothendieck, Catégories Cofibrées Additives et Complexe Cotangent Relatif. IV, 167 pages. 1968. DM 12,–

Vol. 80: Seminar on Triples and Categorical Homology Theory. Edited by B. Eckmann. IV, 398 pages. 1969. DM 20,–

Vol. 81: J.-P. Eckmann et M. Guenin, Méthodes Algébriques en Mécanique Statistique. VI, 131 pages. 1969. DM 12,–

Vol. 82: J. Wloka, Grundräume und verallgemeinerte Funktionen. VIII, 131 Seiten. 1969. DM 12,–

Vol. 83: O. Zariski, An Introduction to the Theory of Algebraic Surfaces. IV, 100 pages. 1969. DM 8,–

Vol. 84: H. Lüneburg, Transitive Erweiterungen endlicher Permutationsgruppen. IV, 119 Seiten. 1969. DM 10,–

Vol. 85: P. Cartier et D. Foata, Problèmes combinatoires de commutation et réarrangements. IV, 88 pages. 1969. DM 8,–

Vol. 86: Category Theory, Homology Theory and their Applications I. Edited by P. Hilton. VI, 216 pages. 1969. DM 16,–

Vol. 87: M. Tierney, Categorical Constructions in Stable Homotopy Theory. IV, 65 pages. 1969. DM 6,–

Vol. 88: Séminaire de Probabilités III. IV, 229 pages. 1969. DM 18,–

Vol. 89: Probability and Information Theory. Edited by M. Behara, K. Krickeberg and J. Wolfowitz. IV, 256 pages. 1969. DM 18,–

Vol. 90: N. P. Bhatia and O. Hajek, Local Semi-Dynamical Systems. II, 157 pages. 1969. DM 14,–

Vol. 91: N. N. Janenko, Die Zwischenschrittmethode zur Lösung mehrdimensionaler Probleme der mathematischen Physik. VIII, 194 Seiten. 1969. DM 16,80

Vol. 92: Category Theory, Homology Theory and their Applications II. Edited by P. Hilton. V, 308 pages. 1969. DM 20,–

Vol. 93: K. R. Parthasarathy, Multipliers on Locally Compact Groups. III, 54 pages. 1969. DM 5,60

Vol. 94: M. Machover and J. Hirschfeld, Lectures on Non-Standard Analysis. VI, 79 pages. 1969. DM 6,–

Vol. 95: A. S. Troelstra, Principles of Intuitionism. II, 111 pages. 1969. DM 10,–

Vol. 96: H.-B. Brinkmann und D. Puppe, Abelsche und exakte Kategorien, Korrespondenzen. V, 141 Seiten. 1969. DM 10,–

Vol. 97: S. O. Chase and M. E. Sweedler, Hopf Algebras and Galois theory. II, 133 pages. 1969. DM 10,–

Vol. 98: M. Heins, Hardy Classes on Riemann Surfaces. III, 106 pages. 1969. DM 10,–

Vol. 99: Category Theory, Homology Theory and their Applications III. Edited by P. Hilton. IV, 489 pages. 1969. DM 24,–

Vol. 100: M. Artin and B. Mazur, Etale Homotopy. II, 196 Seiten. 1969. DM 12,–

Vol. 101: G. P. Szegö et G. Treccani, Semigruppi di Trasformazioni Multivoche. VI, 177 pages. 1969. DM 14,–

Vol. 102: F. Stummel, Rand- und Eigenwertaufgaben in Sobolewschen Räumen. VIII, 386 Seiten. 1969. DM 20,–

Vol. 103: Lectures in Modern Analysis and Applications I. Edited by C. T. Taam. VII, 162 pages. 1969. DM 12,–

Vol. 104: G. H. Pimbley, Jr., Eigenfunction Branches of Nonlinear Operators and their Bifurcations. II, 128 pages. 1969. DM 10,–

Vol. 105: R. Larsen, The Multiplier Problem. VII, 284 pages. 1969. DM 18,–

Vol. 106: Reports of the Midwest Category Seminar III. Edited by S. Mac Lane. III, 247 pages. 1969. DM 16,–

Vol. 107: A. Peyerimhoff, Lectures on Summability. III, 111 pages. 1969. DM 8,–

Vol. 108: Algebraic K-Theory and its Geometric Applications. Edited by R. M. F. Moss and C. B. Thomas. IV, 86 pages. 1969. DM 6,–

Vol. 109: Conference on the Numerical Solution of Differential Equations. Edited by J. Ll. Morris. VI, 275 pages. 1969. DM 18,–

Vol. 110: The Many Facets of Graph Theory. Edited by G. Chartrand and S. F. Kapoor. VIII, 290 pages. 1969. DM 18,–

Vol. 111: K. H. Mayer, Relationen zwischen charakteristischen Zahlen. III, 99 Seiten. 1969. DM 8,-

Vol. 112: Colloquium on Methods of Optimization. Edited by N. N. Moiseev. IV, 293 pages. 1970. DM 18,-

Vol. 113: R. Wille, Kongruenzklassengeometrien. III, 99 Seiten. 1970. DM 8,-

Vol. 114: H. Jacquet and R. P. Langlands, Automorphic Forms on GL (2). VII, 548 pages. 1970. DM 24,-

Vol. 115: K. H. Roggenkamp and V. Huber-Dyson, Lattices over Orders I. XIX, 290 pages. 1970. DM 18,-

Vol. 116: Sèminaire Pierre Lelong (Analyse) Année 1969. IV, 195 pages. 1970. DM 14,-

Vol. 117: Y. Meyer, Nombres de Pisot, Nombres de Salem et Analyse Harmonique. 63 pages. 1970. DM 6,-

Vol. 118: Proceedings of the 15th Scandinavian Congress, Oslo 1968. Edited by K. E. Aubert and W. Ljunggren. IV, 162 pages. 1970. DM 12,-

Vol. 119: M. Raynaud, Faisceaux amples sur les schémas en groupes et les espaces homogènes. III, 219 pages. 1970. DM 14,-

Vol. 120: D. Siefkes, Büchi's Monadic Second Order Successor Arithmetic. XII, 130 Seiten. 1970. DM 12,-

Vol. 121: H. S. Bear, Lectures on Gleason Parts. III, 47 pages. 1970. DM 6,-

Vol. 122: H. Zieschang, E. Vogt und H.-D. Coldewey, Flächen und ebene diskontinuierliche Gruppen. VIII, 203 Seiten. 1970. DM 16,-

Vol. 123: A. V. Jategaonkar, Left Principal Ideal Rings. VI, 145 pages. 1970. DM 12,-

Vol. 124: Séminare de Probabilités IV. Edited by P. A. Meyer. IV, 282 pages. 1970. DM 20,-

Vol. 125: Symposium on Automatic Demonstration. V, 310 pages. 1970. DM 20,-

Vol. 126: P. Schapira, Théorie des Hyperfonctions. XI, 157 pages. 1970. DM 14,-

Vol. 127: I. Stewart, Lie Algebras. IV, 97 pages. 1970. DM 10,-

Vol. 128: M. Takesaki, Tomita's Theory of Modular Hilbert Algebras and its Applications. II, 123 pages. 1970. DM 10,-

Vol. 129: K. H. Hofmann, The Duality of Compact Semigroups and C*- Bigebras. XII, 142 pages. 1970. DM 14,-

Vol. 130: F. Lorenz, Quadratische Formen über Körpern. II, 77 Seiten. 1970. DM 8,-

Vol. 131: A Borel et al., Seminar on Algebraic Groups and Related Finite Groups. VII, 321 pages. 1970. DM 22,-

Vol. 132: Symposium on Optimization. III, 348 pages. 1970. DM 22,-

Vol. 133: F. Topsøe, Topology and Measure. XIV, 79 pages. 1970. DM 8,-

Vol. 134: L. Smith, Lectures on the Eilenberg-Moore Spectral Sequence. VII, 142 pages. 1970. DM 14,-

Vol. 135: W. Stoll, Value Distribution of Holomorphic Maps into Compact Complex Manifolds. II, 267 pages. 1970. DM 18,-

Vol. 136: M. Karoubi et al., Séminaire Heidelberg-Saarbrücken-Strasbuorg sur la K-Théorie. IV, 264 pages. 1970. DM 18,-

Vol. 137: Reports of the Midwest Category Seminar IV. Edited by S. MacLane. III, 139 pages. 1970. DM 12,-

Vol. 138: D. Foata et M. Schützenberger, Théorie Géométrique des Polynômes Eulériens. V, 94 pages. 1970. DM 10,-

Vol. 139: A. Badrikian, Séminaire sur les Fonctions Aléatoires Linéaires et les Mesures Cylindriques. VII, 221 pages. 1970. DM 18,-

Vol. 140: Lectures in Modern Analysis and Applications II. Edited by C. T. Taam. VI, 119 pages. 1970. DM 10,-

Vol. 141: G. Jameson, Ordered Linear Spaces. XV, 194 pages. 1970. DM 16,-

Vol. 142: K. W. Roggenkamp, Lattices over Orders II. V, 388 pages. 1970. DM 22,-

Vol. 143: K. W. Gruenberg, Cohomological Topics in Group Theory. XIV, 275 pages. 1970. DM 20,-

Vol. 144: Seminar on Differential Equations and Dynamical Systems, II. Edited by J. A. Yorke. VIII, 268 pages. 1970. DM 20,-

Vol. 145: E. J. Dubuc, Kan Extensions in Enriched Category Theory. XVI, 173 pages. 1970. DM 16,-

Vol. 146: A. B. Altman and S. Kleiman, Introduction to Grothendieck Duality Theory. II, 192 pages. 1970. DM 18,-

Vol. 147: D. E. Dobbs, Cech Cohomological Dimensions for Commutative Rings. VI, 176 pages. 1970. DM 16,-

Vol. 148: R. Azencott, Espaces de Poisson des Groupes Localement Compacts. IX, 141 pages. 1970. DM 14,-

Vol. 149: R. G. Swan and E. G. Evans, K-Theory of Finite Groups and Orders. IV, 237 pages. 1970. DM 20,-

Vol. 150: Heyer, Dualität lokalkompakter Gruppen. XIII, 372 Seiten. 1970. DM 20,-

Vol. 151: M. Demazure et A. Grothendieck, Schémas en Groupes I. (SGA 3). XV, 562 pages. 1970. DM 24,-

Vol. 152: M. Demazure et A. Grothendieck, Schémas en Groupes II. (SGA 3). IX, 654 pages. 1970. DM 24,-

Vol. 153: M. Demazure et A. Grothendieck, Schémas en Groupes III. (SGA 3). VIII, 529 pages. 1970. DM 24,-

Vol. 154: A. Lascoux et M. Berger, Variétés Kähleriennes Compactes. VII, 83 pages. 1970. DM 8,-

Vol. 155: Several Complex Variables I, Maryland 1970. Edited by J. Horváth. IV, 214 pages. 1970. DM 18,-

Vol. 156: R. Hartshorne, Ample Subvarieties of Algebraic Varieties. XIV, 256 pages. 1970. DM 20,-

Vol. 157: T. tom Dieck, K. H. Kamps und D. Puppe, Homotopietheorie. VI, 265 Seiten. 1970. DM 20,-

Vol. 158: T. G. Ostrom, Finite Translation Planes. IV. 112 pages. 1970. DM 10,

Vol. 159: R. Ansorge und R. Hass. Konvergenz von Differenzenverfahren für lineare und nichtlineare Anfangswertaufgaben. VIII, 145 Seiten. 1970. DM 14,-

Vol. 160: L. Sucheston, Constributions to Ergodic Theory and Probability. VII, 277 pages. 1970. DM 20,-

Vol. 161: J. Stasheff, H-Spaces from a Homotopy Point of View. VI, 95 pages. 1970. DM 10,-

Vol. 162: Harish-Chandra and van Dijk, Harmonic Analysis on Reductive p-adic Groups. IV, 125 pages. 1970. DM 12,-

Vol. 163: P. Deligne, Equations Différentielles à Points Singuliers Reguliers. III, 133 pages. 1970. DM 12,-

Vol. 164: J. P. Ferrier, Seminaire sur les Algebres Complètes. II, 69 pages. 1970. DM 8,

Vol. 165: J. M. Cohen, Stable Homotopy. V, 194 pages. 1970. DM 16,-

Vol. 166: A. J. Silberger, PGL_2 over the p-adics: its Representations, Spherical Functions, and Fourier Analysis. VII, 202 pages. 1970. DM 18,-

Vol. 167: Lavrentiev, Romanov and Vasiliev, Multidimensional Inverse Problems for Differential Equations. V, 59 pages. 1970. DM 10,-

Vol. 168: F. P. Peterson, The Steenrod Algebra and its Applications: A conference to Celebrate N. E. Steenrod's Sixtieth Birthday. VII, 317 pages. 1970. DM 22,-

Vol. 169: M. Raynaud, Anneaux Locaux Henséliens. V, 129 pages. 1970. DM 12,-

Vol. 170: Lectures in Modern Analysis and Applications III. Edited by C. T. Taam. VI, 213 pages. 1970. DM 18,-

Vol. 171: Set-Valued Mappings, Selections and Topological Properties of 2^X. Edited by W. M. Fleischman. X, 110 pages. 1970. DM 12,-

Vol. 172: Y.-T. Siu and G. Trautmann, Gap-Sheaves and Extension of Coherent Analytic Subsheaves. V, 172 pages. 1971. DM 16,-

Vol. 173: J. N. Mordeson and B. Vinograde, Structure of Arbitrary Purely Inseparable Extension Fields. IV, 138 pages. 1970. DM 14,-

Vol. 174: B. Iversen, Linear Determinants with Applications to the Picard Scheme of a Family of Algebraic Curves. VI, 69 pages. 1970. DM 8,-

Vol. 175: M. Brelot, On Topologies and Boundaries in Potential Theory. VI, 176 pages. 1971. DM 18,-

Vol. 176: H. Popp, Fundamentalgruppen algebraischer Mannigfaltigkeiten. IV, 154 Seiten. 1970. DM 16,-

Vol. 177: J. Lambek, Torsion Theories, Additive Semantics and Rings of Quotients. VI, 94 pages. 1971. DM 12,-

Vol. 178: Th. Bröcker und T. tom Dieck, Kobordismentheorie. XVI, 191 Seiten. 1970. DM 18,-

Vol. 179: Seminaire Bourbaki – vol. 1968/69. Exposés 347-363. IV, 295 pages. 1971. DM 22,-

Vol. 180: Séminaire Bourbaki – vol. 1969/70. Exposés 364-381. IV, 310 pages. 1971. DM 22,-

Vol. 181: F. DeMeyer and E. Ingraham, Separable Algebras over Commutative Rings. V, 157 pages. 1971. DM 16,-